Oxidation and Reduction in Inorganic and Analytical Chemistry

Oxidation and Reduction in Inorganic and Analytical Chemistry

A Programmed Introduction

ALAN VINCENT

*School of Inorganic and Physical Chemistry,
Kingston Polytechnic*

JOHN WILEY & SONS

Chichester . New York · Brisbane · Toronto · Singapore

Library of Congress Cataloging in Publication Data:

Vincent, Alan.
  Oxidation and reduction in inorganic and analytical
chemistry.
  Includes index
  1. Oxidation–reduction reaction—Programmed
instruction. I. Title
QD63.09V57   1985   541.3'93   84–29096

ISBN 0 471 90698 0

British Library Cataloguing in Publication Data:

Vincent. A.
  Oxidation and reduction in inorganic and
  analytical chemistry.
  1. Oxidation–reduction reaction
  I. Title
  541.3'93   QD63.09

ISBN 0 471 90698 0

Typeset by Mathematical Composition Setters Ltd, Salisbury, UK
Printed in Great Britain

Preface

Many reactions in inorganic and analytical chemistry involve oxidation and reduction, but the associated ideas of equation balancing, mole calculations and simple thermodynamics cause difficulties for students well into undergraduate courses.

The present set of programmes has been developed to tackle this problem by providing self-instructional backup for students at GCE Advanced Level and beyond. The aim is to start from the basics of the topic and proceed to material normally covered in a Degree or other post-school course. This broad coverage may tempt some students to omit either the more elementary or the more advanced aspects of the book. Experience suggests, however, that revision of basic ideas and studying a topic in more depth than strictly necessary can both be beneficial.

The first two programmes are basic material, after which the student can study either analytical calculations or the thermodynamic aspects of the topic. The way in which each programme relies on material developed earlier is shown in the diagram below:

Prog. 1. Basic Principles

Prog. 2. Oxidation Numbers and Equations

Prog. 3. Analytical Calculations   Prog. 4. Thermodynamic Aspects

Prog. 5. Oxidation State Diagrams

The programmes were used in their original form at Kingston Polytechnic and at some other Colleges and Schools. The feedback from these trials, together with comments and suggestions from colleagues, has been incorporated into the final version of the programmes and I am grateful for the help given by both staff and students in this validation process.

Kingston Polytechnic   Alan Vincent
          1984

Contents

How to Use the Programmes

Each programme starts with a list of learning objectives, and a summary of the knowledge you will need before starting. You should study these sections carefully and make good any deficiencies in your previous knowledge. You may find it helpful at this stage to look at the revision notes at the end of the programme, which give a summary of the material covered. The tests, also at the end, will show you the sort of problems you should be able to tackle after working through the main text (but do not at this stage look at the answers!).

The body of each programme consists of information presented in small numbered sections termed *frames*. Each frame ends with a problem or question and then a line. You should cover the page with a sheet of paper or card and pull it down until you come to the line at the end of the frame. Read the frame and **write down** your answer to the question. **This is most important—** your learning will be much greater if you commit yourself actively by writing your answer down. You can check immediately whether or not your answer is right because each frame starts with the correct answer to the previous frame's question.

If you work through the whole programme in this way you will be learning at your own pace and checking on your progress as you go. If you are working at about the right pace you should get most of the questions right, but if you get one wrong you should read the frame again, look at the question, its answer and any explanation offered, and try to understand how the answer was obtained. When you are satisfied about the answer, go on to the next frame.

Learning a subject (as opposed to just reading a book about it) can be a long job. Do not get discouraged if you find the programmes taking a long time. Some students find this subject easy and work through each programme in under an hour. Others have been known to take up to two hours for some programmes. Provided the programme objectives are achieved, the time spent is relatively unimportant.

After completing each programme, try the test at the end and be guided in your further study by the notes following the answer section.

Each programme finishes with a page of revision notes, which should be helpful either to summarize the programme before or after use, or to serve as revision material later.

I hope you find the programmes enjoyable and useful.

Programme 1

Basic Principles

Objectives

After completing this programme you should be able to:
1. Recognize an oxidation–reduction reaction and state what is being oxidized and what is being reduced.
2. State concisely why oxidation and reduction always occur together.
3. Explain why addition of oxygen, or any other electronegative element, is similar to electron removal.
4. Explain why addition of hydrogen, or any other electropositive element, is similar to electron addition.

Assumed Knowledge

Before starting the programme, you should have a knowledge of how simple compounds such as CaO and $CaCl_2$ can be written as ions and you should also have a knowledge of whether an element can be classed as electropositive or electronegative.

Basic Principles

1.1 We are all familiar with reactions involving oxygen. Oxygen combines with a great many elements and compounds, so it is convenient to call all such reactions oxidations. Some examples are:

$$2Cu + O_2 \rightarrow 2CuO \qquad (1.1)$$

$$2H_2 + O_2 \rightarrow 2H_2O \qquad (1.2)$$

Write the equation for the oxidation of calcium by oxygen.

1.2 $$2Ca + O_2 \rightarrow 2CaO \qquad (1.3)$$

The reverse process, the removal of oxygen, is termed reduction, and can often be performed by using hydrogen to remove the oxygen:

$$CuO + H_2 \rightarrow Cu + H_2O \qquad (1.4)$$

Compare this equation with equation 1.2. What has happened to the hydrogen?

1.3 It has been oxidized to water.

This illustrates a very important point, namely that oxidation and reduction must occur together. The following reaction is an oxidation–reduction (redox) reaction. State the substance which is oxidized and the substance which is reduced.

$$Ca + H_2O \rightarrow CaO + H_2 \qquad (1.5)$$

1.4 Calcium is oxidized to its oxide, water is reduced to hydrogen.

In the five equations so far, we have used seven chemical substances, namely

$$H_2 \quad Ca \quad Cu \quad O_2 \quad H_2O \quad CaO \quad CuO$$

Some of these could possibly act as oxidizing agents by giving oxygen to some other element or compound. Which are the possible oxidising agents in the list?

1.5 O_2, H_2O, CaO, CuO. Oxygen and all the compounds containing oxygen could possibly give it up to some other substance. The other three substances, namely H_2, Ca and Cu, can only accept oxygen from some suitable donor. They therefore react in a manner opposite to oxidizing agents and can be called...

1.6   Reducing agents.

We now have a list of four oxidizing agents and three reducing agents, and it is reasonable to ask which is the strongest oxidizing agent, which is the next strongest, and so on.

Which of the four oxidizing agents will oxidize any of the three reducing agents?

1.7 O_2. It is therefore the strongest of these oxidizing agents.
Now let us look again at equation 1.4:

$$CuO + H_2 \rightarrow Cu + H_2O \qquad (1.4)$$

Here we have a competition between copper oxide and water to give up oxygen and cause an oxidation reaction.

Which one gives up its oxygen and is therefore the stronger oxidant?

1.8   CuO.

We can therefore say that in order of oxidizing power:

$$O_2 > CuO > H_2O$$

Look again at equation 1.5:

$$Ca + H_2O \rightarrow CaO + H_2 \qquad (1.5)$$

We can again regard this as a competition between two possible oxidizing agents, H_2O and CaO, to give up their oxygen.

Which one actually gives it up and is therefore the stronger oxidizing agent?

1.9 H_2O.

Where then must CaO fit in our series of oxidizing agents placed in order of oxidizing power?

$$O_2 > CuO > H_2O$$

1.10   CaO is the weakest oxidizing agent of all:

$$O_2 > CuO > H_2O > CaO$$

In fact, there is no reaction we have seen so far in which it gives up its oxygen.

In our series of oxidizing agents, copper oxide is quite a powerful oxidizing agent. Remembering this, what reaction do you think could happen if copper oxide and calcium were heater together?

1.11 $$CuO + Ca \rightarrow CaO + Cu \qquad (1.6)$$

as copper oxide is a stronger oxidizing agent than calcium oxide.

In the same way as we compared the strengths of oxidizing agents, we can compare reducing agents.

Look back at equations 1.4–1.6. Which of the three reducing agents, H_2, Cu and Ca, will remove oxygen from either of the other two?

1.12   Ca.

Calcium must therefore be the strongest reducing agent.
In general, if a metal oxide is a poor oxidizing agent (e.g. CaO), the metal will be a good reducing agent (e.g. Ca), as a metal which is good at attracting oxygen will be poor at releasing it again.

Use a similar reasoning (with equations 1.4–1.6) to put hydrogen and copper with calcium in order of strength as reducing agents.

1.13   (Strongest) $Ca > H_2 > Cu$ (weakest)

Calcium is able to reduce water to hydrogen (equation 1.5) and hydrogen reduces copper oxide to copper (equation 1.4).

We shall now consider what is happening in electronic terms when an element is oxidized or reduced. The simplest way of doing this is to consider compounds to be ionic. Although this is not always strictly accurate, it does enable us to relate oxidation and reduction to electronic changes.

Write calcium oxide in ionic form.

1.14 $Ca^{2+}\ O^{2-}$

Thus when calcium is oxidized by oxygen it changes from the metal to the 2+ ion. What has to be done to the calcium (in terms of electrons) to effect this change?

1.15   Two electrons have to be removed: $Ca \rightarrow Ca^{2+} + 2e^-$

Write the result of reacting calcium with Cl_2, F_2 and S in ionic form.

1.16 $Ca^{2+}\quad Cl^-\quad Cl^-$
                         $Ca^{2+}\quad F^-\quad\ \ F^-$
                         $Ca^{2+}\quad S^{2-}$

The reaction of calcium with chlorine is therefore similar to its reaction with oxygen or any other electronegative element. They all cause electrons to be removed and the metal to become a positive ion. Because of this similarity, we can classify all such reactions under the heading of oxidations and define oxidation as follows:

                Oxidation is electron removal.

Define reduction in electronic terms.

1.17   Reduction is electron addition, e.g.

                       $2e^- + S \rightarrow S^{2-}$
                       $2e^- + O \rightarrow O^{2-}$

These equations are called **ION-ELECTRON HALF EQUATIONS**

Complete the following ion-electron half equations and state if they are oxidations or reductions:

                     $e^- + Fe^{3+} \rightarrow \ldots$
                     $2e^- + I_2 \rightarrow \ldots$
                     $Cu^+ \rightarrow \ldots + e^-$
                     $Ce^{3+} \rightarrow Ce^{4+} + \ldots$

1.18

$$e^- + Fe^{3+} \rightarrow Fe^{2+} \qquad \text{reduction}$$
$$2e^- + I_2 \rightarrow 2I^- \qquad \text{reduction}$$
$$Cu^+ \rightarrow Cu^{2+} + e^- \qquad \text{oxidation}$$
$$Ce^{3+} \rightarrow Ce^{4+} + e^- \qquad \text{oxidation}$$

Real chemical reactions (other than those occurring at cell electrodes) do not involve a net addition or removal of electrons, so to obtain an equation for a complete reaction we must add together two of these ion-electron half equations, one for an oxidation and one for a reduction, so that the electrons cancel out, e.g.

$$Ce^{4+} + e^- \rightarrow Ce^{3+} \qquad \text{reduction}$$
$$\underline{Fe^{2+} \rightarrow Fe^{3+} + e^-} \qquad \text{oxidation}$$
$$Fe^{2+} + Ce^{4+} \rightarrow Fe^{3+} + Ce^{3+} \qquad \text{net reaction}$$

Try writing the ion-electron half equations for:
  a. The oxidation of iodide to I_2.
  b. The reduction of Cl_2 to chloride.

1.19

$$\text{a. } 2I^- \rightarrow I_2 + 2e^-$$
$$\text{b. } Cl_2 + 2e^- \rightarrow 2Cl^-$$

Now add these two equations together to obtain the complete equation for the oxidation of iodide ions to iodine by chlorine gas.

1.20

$$2e^- + Cl_2 \rightarrow 2Cl^- \qquad \text{reduction}$$
$$\underline{2I^- \rightarrow I_2 + 2e^-} \qquad \text{oxidation}$$
$$2I^- + Cl_2 \rightarrow I_2 + 2Cl^- \qquad \text{net reaction}$$

These examples are relatively easy because the same number of electrons is involved in both the oxidation and the reduction. In some cases, however, one or both half equations has to be multiplied through by a number to balance out the number of electrons, e.g.

Obtain the net reaction for the following:

$$Ce^{4+} + e^- \rightarrow Ce^{3+} \qquad \text{(reduction of cerium)}$$
$$C_2O_4^{2-} \rightarrow 2CO_2 + 2e^- \qquad \text{(oxidation of ethanedioate)}$$

Hint: multiply one equation all through so that both equations involve the same number of electrons, then add them.

1.21   Multiply the first equation by 2: $2Ce^{4+} + 2e^- \rightarrow 2Ce^{3+}$

Add the second equation: $C_2O_4^{2-} \rightarrow 2e^- + 2CO_2$

$$2Ce^{4+} + C_2O_4^{2-} \rightarrow 2Ce^{3+} + 2CO_2$$

Do the same thing with the following two half equations. Obtain the same number of electrons on each side and then add the resulting equations to obtain the overall equation for the reaction:

$$MnO_4^- + 5e^- + 8H^+ \rightarrow Mn^{2+} + 4H_2O$$

$$C_2O_4^{2-} \rightarrow 2CO_2 + 2e^-$$

1.22   Multiply the first equation by 2:

$$2MnO_4^- + 10e^- + 16H^+ \rightarrow 2Mn^{2+} + 8H_2O$$

Multiply the second equation by 5:

$$5C_2O_4^{2-} \rightarrow 10CO_2 + 10e^-$$

Add: $2MnO_4^- + 5C_2O_4^{2-} + 16H^+ \rightarrow 2Mn^{2+} + 8H_2O + 10CO_2\uparrow$

This equation between manganate(VII) and ethanedioate is very difficult to balance by trial and error, but becomes relatively trivial when the oxidation and reduction processes are split into their two ion-electron half equations. The problem now is how do we obtain a fairly complicated half equation such as that for the reduction of manganate(VII)? This is best done by using the oxidation number method, which is the subject of Programme 2.

Basic Principles Test

1. Which of the following are oxidation–reduction reactions?:

    i. $Zn + Br_2 \rightarrow ZnBr_2$

    ii. $Fe + 2H^+ \rightarrow Fe^{2+} + H_2\uparrow$

    iii. $CaCO_3 \rightarrow CaO + CO_2$

    iv. $N_2H_4 + 2I_2 + 4OH^- \rightarrow 4I^- + N_2 + 4H_2O$

2. State, for each of the oxidation–reduction reactions above, what is oxidized and what is reduced.

Answers

			Marks
1.	i, ii and iv are oxidation reduction reactions		3
2.	i.	Zn is oxidized to Zn^{2+}	1
		Br_2 is reduced to Br^-	1
	ii.	Fe is oxidized to Fe^{2+}	1
		H^+ is reduced to H_2	1
	iv.	N_2H_4 is oxidized to N_2	1
		I_2 is reduced to I^-	1
		Total	9

To be able to proceed confidently to the next programme, you should have obtained at least 6 marks on this test. If you have not achieved this level, look back at Frames 1.13−1.22 and try to understand the answers given for the test. Programme 2 covers some similar material in a slightly different way, so it may be that this test will become clearer after that programme.

Basic Principles

Revision Notes

Oxidation and reduction always occur together—one element or compound being oxidized by another which is itself reduced.

We can speak of the relative oxidizing or reducing power of different substances; for example water is a stronger oxidant than calcium oxide, so water oxidizes calcium. Water, however, is a weaker oxidant than copper oxide, so copper oxide oxidizes hydrogen to water.

Oxidation can always be represented as electron removal and reduction as electron gain. The two processes can therefore be expressed by ion-electron half reactions which show electron loss or gain explicity.

Programme 2

Oxidation Numbers and Equations

Objectives

After completing this programme you should be able to:
1. Assign oxidation numbers to simple species.
2. Write ion-electron half equations for simple oxidations and reductions.
3. Combine ion-electron half equations to give net reactions.

Assumed Knowledge

Before starting the programme you should have some knowledge of the outer electron structure of Main Group elements (such as Na, O, Cl, Ca, N, S and C), and you should also be familiar with the material in Programme 1.

Introduction

The oxidation number of an atom in a chemical substance is a convenient way of expressing, with some degree of physical reality, the extent to which the atom has been oxidized (relative to the free element) in forming the compound. The oxidation number of an element is always unambiguously defined by a set of rules, but the degree of physical reality to be ascribed to an oxidation number is limited. Nevertheless, the idea is a convenient book-keeping concept which allows equations to be constructed with great ease.

The oxidation number rules are as follows:
  i.   The oxidation number (or state) of an atom in an element is always zero.
  ii. The oxidation number of an ionised atom, e.g. Na^+ or Cl^-, is equal to the charge on the ion (including the sign).
  iii. The oxidation number of an atom in a compound, or in a complex ion, is equal to the charge the atom would have in the most likely ionic formulation of the compound. E.g. oxygen is commonly -2, hydrogen $+1$, halogens -1, Group I metals (Na, etc.) $+1$, Group II metals (Ca, etc.) $+2$.
  iv. The algebraic sum of the oxidation numbers of all atoms in a complex ion equals the ionic charge. (*N.B.* a 'complex ion' includes any multi-atom ion such as NO_3^-).

The only rule which is at all difficult to apply in practice is iii, which requires some judgement of the 'most likely ionic formulation.' This is not usually too much trouble, however, especially if one starts from a presumption that the more electronegative atom will probably assume a closed shell negative ion structure, e.g. SO_4^{2-} would become $S^{6+} + 4O^{2-}$ and the oxidation numbers would be assigned as $+6$ and -2.

The first few frames are practice at working out oxidation numbers.

Oxidation Numbers and Equations

2.1 Assign oxidation numbers to the following (rules i and ii):

Na^+.   Oxidation no. of Na   = $+1$
O^{2-}.   Oxidation no. of O   = -2
Cl^-.   Oxidation no. of Cl   =
Ca^{2+}.   Oxidation no. of Ca   =
O_2.   Oxidation no. of O   =
C (diamond) Oxidation no. of C   =
C (graphite). Oxidation no. of C   =

2.2 Cl^-: -1
             Ca^{2+}: $+2$
             O_2:   zero
             C: zero (independent of allotropic form)

Let us now turn to rule iii. Some compounds are easy:

$$AlCl_3: \quad Al^{3+} \quad 3Cl^-$$
Oxidation numbers: Al $+3$, Cl -1

In others the 'most likely ionic formulation' is less likely, e.g. MnO_4^- becomes $Mn^{7+} + 4O^{2-}$, so manganese is in the $+7$ oxidation state and oxygen is in the -2 state. Confirm rule iv for MnO_4^-.

2.3 Algebraic sum $= +7 + (4 \times -2) = -1$, which is the charge on the original ion.

Assign oxidation numbers to Cr and O in the chromate ion, CrO_4^{2-}.

2.4 CrO_4^{2-} becomes $Cr^{6+} + 4O^{2-}$, hence the oxidation numbers are Cr $+6$ and O -2

There is no suggestion in this formalism that the metals actually have these very high charges. The method is simply a convenient book-keeping device to enable equations to be balanced easily and oxidation reduction processes to be understood.

Assign oxidation numbers to chromium and oxygen in the dichromate ion, $Cr_2O_7^{2-}$

2.5 $Cr_2O_7^{2-}$ become $Cr^{6+} + Cr^{6+} + 7O^{2-}$, hence the oxidation numbers are Cr $+6$ and O -2, i.e. the same as in CrO_4^{2-}.

Let us now return to the half equation for the reduction of manganate(VII) (containing manganese in the $+7$ state) to Mn^{2+}.

Complete the following equation by adding electrons:

$$Mn^{7+} \rightarrow Mn^{2+}$$

2.6 $Mn^{7+} + 5e^- \rightarrow Mn^{2+}$ (2.1)

This is a reduction of the manganese because the oxidation number (or oxidation state) has decreased by 5 units, so 5 electrons are needed.

Manganese cannot really exist as the ion Mn^{7+}; in this oxidation state it exists as MnO_4^-. In an aqueous solution of MnO_4^- there is an abundant supply of water, which we can write as:

$$H_2O \rightleftharpoons H^+(aq) + OH^-(aq)$$

Work out the oxidation numbers of hydrogen and oxygen in H_2O, H^+ and OH^-.

2.7 H_2O: H $+1$, O -2
                H^+:   H $+1$
                OH^-:   H $+1$, O -2

That is, the self-ionisation of water does not involve any oxidation or reduction. We can therefore add H_2O, H^+ or OH^- to an equation representing an aqueous reaction without in any way influencing our conclusions about redox processes. Let us now go back to equation 2.1:

$$Mn^{7+} + 5e^- \rightarrow Mn^{2+} \qquad (2.1)$$

If we re-write Mn^{7+} as MnO_4^- and add H^+ and H_2O, we obtain

$$MnO_4^- + 5e^- + 8H^+ \rightarrow Mn^{2+} + 4H_2O$$

Add up the ionic charges on both sides of this equation.

2.8   Left-hand side: -1 -5 $+8$ $= +2$
                          Right-hand side: $+2$

i.e. the charges balance, as they must in any equation, and we have constructed an ion-electron half equation starting from the initial and final oxidation states of manganese.

We shall now try to do the same thing for sulphur in SO_4^{2-} and SO_3^{2-}.

Work out the oxidation number of sulphur in SO_4^{2-} and SO_3^{3-}

2.9 $SO_4^{2-} \equiv S^{6+} + 4O^{2-}$ sulphur is in oxidation state $+6$
                 $SO_3^{2-} \equiv S^{4+} + 3O^{2-}$ sulphur is in oxidation state $+4$

Note that in the sulphate(IV) case (and in many others) it is not possible to have a closed shell configuration for all ions. In these cases the more electronegative atom (in this case oxygen) is given a closed shell configuration.

The oxidation of sulphate(IV) to sulphate(VI) can therefore be represented simply as an increase in the oxidation number of the sulphur:

$$S^{4+} \rightarrow S^{6+} + 2e^- \tag{2.2}$$

or

$$SO_3^{2-} \rightarrow SO_4^{2-} + 2e^- \tag{2.3}$$

Now add H_2O and H^+ to equation 2.3 to make an ion-electron half equation describing the oxidation of sulphate(IV) to sulphate(VI) in acidic solution.

Hint: the equation requires oxygen (from H_2O) on the left-hand side.

2.10 $$SO_3^{2-} + H_2O \rightarrow SO_4^{2-} + 2H^+ + 2e^- \tag{2.4}$$

Combine this equation (by balancing electrons as in Frames 1.18–1.22) with the equation for the reduction of chlorine to chloride and hence obtain the equation for the oxidation of sulphate(IV) to sulphate(VI) by chlorine.

2.11

$$SO_3^{2-} + H_2O \rightarrow SO_4^{2-} + 2H^+ + 2e^-$$
$$2e^- + Cl_2 \rightarrow 2Cl^-$$

$$SO_3^{2-} + Cl_2 + H_2O \rightarrow SO_4^{2-} + 2H^+ + 2Cl^-$$

Let us now turn to a different oxidising agent, potassium dichromate(VI), $K_2Cr_2O_7$. We have already seen in Frame 2.5 that chromium has an oxidation state of $+6$ in this compound. When it acts as an oxidizing agent, the chromium goes down to the $+3$ state:

$$2Cr^{6+} \rightarrow 2Cr^{3+}$$

Balance this equation by adding electrons.

2.12

$$2Cr^{6+} + 6e^- \rightarrow 2Cr^{3+}$$

There are two chromium atoms in this because there are two chromium atoms in $Cr_2O_7^{2-}$.

Now try to construct the whole ion-electron half equation by adding H^+ and H_2O to the basic outline equation:

$$Cr_2O_7^{2-} + 6e^- \rightarrow 2Cr^{3+}$$

Hint: the oxygen on the left-hand side must not change its oxidation state.

2.13

$$Cr_2O_7^{2-} + 14H^+ + 6e^- \rightarrow 2Cr^{3+} + 7H_2O \tag{2.5}$$

Now combine this equation (by eliminating electrons as before) with the equation for the oxidation of ethanedioate ($C_2O_4^{2-}$) to carbon dioxide:

$$C_2O_4^{2-} \rightarrow 2CO_2 + 2e^- \tag{2.6}$$

2.14   Multiply equation 2.6 by 3:

$$3C_2O_4^{2-} \rightarrow 6CO_2 + 6e^-$$

Add equation 2.5:

$$Cr_2O_7^{2-} + 14H^+ + 6e^- \rightarrow 2Cr^{3+} + 7H_2O$$

$$3C_2O_4^{2-} + Cr_2O_7^{2-} + 14H^+ \rightarrow 2Cr^{3+} + 7H_2O + 6CO_2$$

This represents the complete reaction between the two substances which, of course, must not involve addition or removal of electrons from the system. The final equation is also balanced both for numbers of atoms and for electrical charges.

Confirm that the equation is balanced for both atoms and charges.

2.15   Atoms: both sides have 6C, 19O, 2Cr and 14H.
Charges: left-hand side; 3×-2, -2, $+14 = +6$
         right-hand side: $2 \times +3$ $= +6$

It is possible to proceed directly from oxidation state changes to a balanced equation without using the intermediate ion-electron half equation. We must simply remember that oxidation and reduction always occur together and the number of electrons involved in either process is the product of the number of atoms and the change in oxidation state. Thus for the reduction of dichromate(VI) to chromium(III):

$$Cr_2O_7^{2-} \rightarrow 2Cr^{3+}$$

There are two atoms of chromium, each decreasing its oxidation state by 3 ($+6$ to $+3$). Hence:

Number of electrons required $= 2 \times 3 = 6$.

How many electrons are released when sulphate(IV) is oxidized to sulphate(VI)?

2.16   Two, i.e. one atom increasing its oxidation state by two units. Thus to balance an equation between dichromate(VI) and sulphate(IV), we have to make the number of electrons equal.

By what factor must we multiply the sulphate equation to achieve this?

2.17   Three, i.e. $\qquad\qquad$ $3S^{IV} \rightarrow 3S^{VI}$

Three atoms, each increasing its oxidation state by 2. Hence

$$\text{Number of electrons} = 3 \times 2 = 6.$$

By this argument we can write down a 'skeleton equation' showing that one dichromate(VI) ion reacts with three sulphate(IV) ions:

reduction, $2 \times 3 = 6e^-$

$$Cr_2O_7^{2-} \quad + \quad 3SO_3^{2-} \rightarrow 3SO_4^{2-} + 2Cr^{3+} \qquad (2.7)$$

oxidation, $3 \times 2 = 6e^-$

We can again balance this equation completely by adding H_2O and H^+ (the reaction proceeds in acidic solution), but we must keep the $1:3$ ratio of principal reactants already shown.

Do this for equation 2.7.

Hint: the right-hand side has four fewer oxygen atoms than the left-hand side. These must be put in as water molecules.

2.18 $$8H^+ + Cr_2O_7^{2-} + 3SO_3^{2-} \rightarrow 3SO_4^{2-} + 2Cr^{3+} + 4H_2O$$

Check that the resulting equation is balanced for electrical charges.

2.19   The algebraic sum of the charges is zero on both sides.

Let us use the oxidation number method to balance the equation for the oxidation of iron(II) to iron(III) by manganate(VII), which is reduced to manganese(II).

How many electrons are needed to reduce manganate(VII)?

2.20   Five: $\qquad\qquad$ $MnO_4^- \rightarrow Mn^{2+}$

Decrease of oxidation state $= 5$.

How many electrons are released when iron(II) is oxidized?

2.21   One: $\qquad\qquad$ $Fe^{II} \rightarrow Fe^{III}$

Increase of oxidation state $= 1$.

Hence write out the 'skeleton equation' (as in Frame 2.17) by balancing out the numbers of electrons.

2.22

$$5e^-$$

$$MnO_4^- + 5Fe^{2+} \rightarrow 5Fe^{3+} + Mn^{2+}$$

$$5e^-$$

Keeping the $1:5$ ratio unchanged, add H^+ and H_2O to this equation to balance it completely.

2.23 $$8H^+ + MnO_4^- + 5Fe^{2+} \rightarrow 5Fe^{3+} + Mn^{2+} + 4H_2O$$

Check that the equation is balanced for electrical charges.

2.24 Left-hand side: $8 - 1 + (5 \times 2) = +17$
Right-hand side: $(5 \times 3) + 2 = +17$

We can thus balance oxidation–reduction equations either by finding the number of electrons explicitly, as in Frame 2.14, or by inferring it from the oxidation state change, as in Frames 2.17 and 2.22. There are some cases in which the former method is easier, for example the oxidation of

ethanedioate: $C_2O_4^{2-} \rightarrow 2CO_2 + 2e^-$

peroxide: $O_2^{2-} \rightarrow O_2 + 2e^-$

thiosulphate(VI): $2S_2O_3^{2-} \rightarrow S_4O_6^{2-} + 2e^-$

In all these cases, the oxidation number method would work but it is much easier to see the number of electrons involved from the ion-electron half equation. It is a matter of experience to be able to decide the easier method in any particular case.

Obtain balanced equations for:

i. The oxidation of hydrogen peroxide to oxygen by manganate(VII).
ii. The oxidation of thiosulphate(VI) to tetrathionate by iodine.

2.25 i. $6H^+ + 2MnO_4^- + 5H_2O_2 \rightarrow 2Mn^{2+} + 5O_2 + 8H_2O$

ii. $2S_2O_3^{2-} + I_2 \rightarrow S_4O_6^{2-} + 2I^-$

Oxidation Numbers and Equations Test

1. Assign oxidation numbers to the atoms in italics;

$$K_2O \qquad SO_2 \qquad PF_5 \qquad NH_3 \qquad NF_3 \qquad MnO_4^{2-} \qquad H_3PO_3$$

2. Write ion-electron half equations for:

   i. The reduction of H_2O_2 to water.
   ii. The reduction of iodate(V) (IO_3^-) to iodide.
   iii. The oxidation of sulphate(VI) to peroxodisulphate(VI) ($S_2O_8^{2-}$).
   iv. The oxidation of NO_2^- to NO_3^-.

3. Write balanced equations for:

   i. The oxidation of iron(II) to iron(III) by peroxodisulphate(VI) ($S_2O_8^{2-}$).
   ii. The reduction of dichromate(VI) by V^{2+} (forming VO^{2+}).
   iii. The oxidation of As_2O_3 to AsO_3^- by manganate(VII) in acidic solution.

Answers

Marks

1. K_2O -2 1
   SO_2 $+4$ 1
   PF_5 $+5$ 1
   NH_3 -3 1
   NF_3 $+3$ 1
   MnO_4^{2-} $+6$ 1
   H_3PO_3 $+3$ 1

2. $H_2O_2 + 2e^- + 2H^+ \rightarrow 2H_2O$ 1
   $IO_3^- + 6e^- + 6H^+ \rightarrow I^- + 3H_2O$ 1
   $2SO_4^{2-} \rightarrow S_2O_8^{2-} + 2e^-$ 1
   $NO_2^- + H_2O \rightarrow NO_3^- + 2e^- + 2H^+$ 1

3. i. $Fe(II) \rightarrow Fe(III)$ 1 electron
        $2e^- + S_2O_8^{2-} \rightarrow 2SO_4^{2-}$ 2 electrons

        $2Fe^{2+} + S_2O_8^{2-} \rightarrow 2SO_4^{2-} + 2Fe^{3+}$ 3

   ii. $V(II) \rightarrow V(IV)$ 2 electrons
        $Cr^{VI}_2O_7^{2-} \rightarrow 2Cr^{3+}$ $2 \times 3 = 6$ electrons

        $3V^{2+} + Cr_2O_7^{2-} + 8H^+ \rightarrow 2Cr^{3+} + 3VO^{2+} + 4H_2O$ 3

   iii. $As^{III}_2O_3 \rightarrow 2As^V O_3^-$ $2 \times 2 = 4$ electrons
        $Mn^{VII}O_4^- \rightarrow Mn^{2+}$ 5 electrons

        $5As_2O_3 + 4MnO_4^- + 2H^+ \rightarrow 10AsO_3^- +$
        $+ 4Mn^{2+} + H_2O$ 3

Total   20

Balanced equations are essential in order to solve analytical calculation problems. These are the subject of the next programme, so do not go on to this until you are really sure that you can construct balanced equations using the methods in this programme. A test score of at least 14 is suggested, but correct answers to question 3 are most important.

Oxidation Numbers and Equations

Revision Notes

All elements in a compound can be assigned oxidation numbers (which may be positive or negative). These are the charges on the elements when they are completely separated as closed shell ions. If it is not possible to make all elements into closed shell ions, the most electronegative elements are given closed shell structures. Free elements (in any allotropic form) are given the oxidation number zero.

Equations can be balanced by equating the number of electrons given to the oxidizing agent and the number removed from the reducing agent. This can be achieved by:

i. Adding together ion-electron half equations so that the electrons cancel out.
ii. Using oxidation numbers and the relationship:

Number of electrons = number of atoms × change of oxidation state

Programme 3

Analytical Calculations

Objectives

After completing this programme you should be able to:
1. Calculate the amount of a reagent in moles given:
    a. the mass and relative molecular mass or
    b. the volume and concentration of a solution.
2. Calculate the mass of an element, given:
    a. the mass of a compound of known formula or
    b. the volume and concentration of a solution.
3. Solve numerical problems based on gravimetric analysis.
4. Solve numerical problems based on volumetric analysis requiring as the answer:
    a. % by weight of a particular substance in a sample or
    b. the equation for a particular reaction.
5. Solve numerical problems based on volumetric analysis, including a back-titration.

All five objectives are tested at the end of the programme.

Assumed Knowledge

Before starting the programme you should be familiar with the concept of the mole and with oxidation numbers. It will be a considerable help if you are adept at balancing oxidation–reduction equations.

Analytical Calculations

3.1 Many topics in analytical chemistry (and some examination questions!) are concerned with calculations. We must be able to use analytical data such as titration figures and weighings to obtain results in any given form, e.g. % of a particular element or concentration of a solution in mol dm^{-3} or in g dm^{-3}. All these calculations are very simple, provided we remember a few basic relationships concerned with the ways in which chemists measure amounts and concentrations. We shall start by revising the measurement of amount of substance using the unit called the mole.

The mole is the amount of substance which contains as many elementary entities as there are atoms in 0.012 kg of carbon-12.

Since the relative atomic masses of carbon and aluminium are 12 and 27, respectively, there are the same number of atoms (elementary entities) in 27 g of aluminium as there are in 12 g (0.012 kg) of carbon. These amounts therefore constitute one mole of each element.

How many moles are there in 54 g of Al?

3.2 Two, since $54 = 2 \times 27$.

Write down a relationship between (a) the mass of 1 mol, (b) the mass and (c) the number of moles, for any substance.

3.3   Number of moles $= \dfrac{\text{mass}}{\text{mass of 1 mol}}$

or mass = number of moles × mass of 1 mol (substitute 'tennis ball' for 'mole' in this last statement and it should become very obvious). The term 'number of moles' is not a particularly acceptable one since we do not refer to 'number of metres' or 'number of grams'. It is always preferable to refer to the physical quantity being considered, such as length, mass or, in this case, amount.
We can therefore write:

$$\text{amount} = \dfrac{\text{mass}}{\text{relative molecular mass (RMM)}}$$

or, including units:

$$\text{amount/mole} = \dfrac{\text{mass/g}}{\text{relative molecular mass/g mol}^{-1}}$$

Always take care to distinguish between the physical quantity being measured (mass, length, amount) and the unit being used to measure it (kilogram, metre, mole).

What is the mass of 2 mol of N (relative atomic mass = 14)?

3.4   28 g.

This answer should have been very easy to obtain, but if the question had asked for the mass of 2 mol of nitrogen, the answer could have been 28 g ($2N = 28$) or 56 g ($2N_2 = 56$). This illustrates that we must be very careful to specify the elementary entity when dealing with the mole. This is always best done by using a chemical symbol, for example:

1 mol of H_2 weighs 2 g,
1 mol of H weighs 1 g,
$\frac{1}{2}$ mol of H_2 weighs 1 g,

but nobody can really tell how much is meant by '1 mol of hydrogen'. Bromine has relative atomic mass of 80. Write down the mass of:

   i.   1 mol of Br.
  ii.   1 mol of Br_2.
 iii.   2 mol of Br.
 iv. $\frac{1}{2}$ mol of Br_2.

3.5   i 80 g;   ii 160 g;   iii 160 g;   iv 80 g.

How many moles of bromine are there in 120 g of bromine?

3.5A   You should be complaining that the question is ambiguous!

Write your answer in two unambiguous ways.

3.6 $1\frac{1}{2}$ mol of Br.
$\frac{3}{4}$ mol of Br_2.

Just to make sure you really know it:

amount (in moles) = ?

3.7   Amount (in moles) = $\dfrac{\text{mass}}{\text{mass of 1 mol (RMM)}}$

or

mass = amount × RMM

Chemists measure amounts in moles for a very simple reason. Balanced chemical equations tell us the amounts of both reactants and products directly in moles. The equation for the burning of ethene is as follows:

$$C_2H_4 + 3O_2 \rightarrow 2CO_2 + 2H_2O \qquad (3.1)$$

This equation tells us that 1 mol of C_2H_4 reacts with 3 mol of O_2 to produce 2 mol of CO_2 and 2 mol of H_2O. We can then convert these amounts to masses using the relationships at the top of this frame.

What mass of water would be produced if 28 g (= 1 mol) of ethene was burnt completely in oxygen? (RMM of $H_2O = 18$).

3.8   36 g. 1 mol (= 28 g) of C_2H_4 produces 2 mol of H_2O when it is burnt. Since 1 mol of H_2O weighs 18 g, 2 mol weigh 36 g.

What mass of oxygen is required to burn completely 28 g (= 1 mol) of C_2H_4? (O = 16).

3.9   96 g. 3 mol of O_2 or 6 mol of O are required and $6 \times 16 = 96$.

Take great care in cases like this when one molecule of a substance contains more than one of the same type of atom.

Gravimetric chemical analysis relies on the masses of substances and simple gravimetric calculations use the relationships already discussed. Problem 3.1 is an easy example of this type of calculation.

Problem 3.1

0.7600 g of material X was analysed gravimetrically for chlorine by converting it to AgCl. After suitable treatment, 0.5000 g of AgCl was obtained.

Calculate % Cl in X.

(Ag = 108; Cl = 35.5; AgCl = 143.5)

$$Ag^+ + Cl^- \rightarrow AgCl \tag{3.2}$$

What amount of AgCl (in moles) was obtained?

3.10 $0.5000/143.5 = 0.00348$ mol.

Look back at equation 3.2. This tells us that 1 mol of Ag^+ reacts with 1 mol of Cl^- to produce 1 mol of AgCl.

What amount of Cl^- (in moles) was present in the sample of X in order to produce 0.00348 mol of AgCl?

3.11   0.00348, since the reaction is $1:1$ by moles.

What is the mass of 0.00348 mol of Cl?

3.12 $0.00348 \times 35.5 = 0.1235$ g.

All this Cl came originally from the sample of 0.7600 g of X.

What is the % Cl in X?

3.13 $\dfrac{0.1235}{0.7600} \times 100 = 16.3\%$.

Problems of this type are generally straightforward because we are dealing directly with masses and amounts, and the amounts can be related directly to the chemical equation. Problems involving solutions can be more tricky because volumes and concentrations have to be converted into amounts before being related to the equation. We must therefore spend a little time now discussing these conversions.

A 1-mol dm^{-3} solution of HCl contains 1 mol (36.5 g) of HCl in 1 dm^3 ($= 1000$ cm^3) of solution. What mass of HCl is contained in:

    i.   500 cm^3
    ii. 100 cm^3
    iii. 25 cm^3 of 1-mol dm^{-3} HCl?

3.14   i. $36.5 \times \dfrac{500}{1000} = 18.25$ g.

    ii. 3.65 g.
    iii. 0.9125 g.

Now let us consider more dilute HCl. What mass of HCl is there in 1 dm^3 of 0.3-mol cm^{-3} HCl?

3.15 $36.5 \times 0.3 = 10.95$ g.
  What mass of HCl is there in:
    i.   500 cm^3
    ii. 100 cm^3
    iii. 25 cm^3 of 0.3-mol dm^{-3} HCl?

3.16   i. $36.5 \times 0.3 \times \dfrac{500}{1000} = 5.475$ g.

    ii. 1.095 g.
    iii. 0.2738 g.

These examples have allowed us to go from a volume and concentration to a mass, but it is as important to be able to calculate an amount of a reagent, so that we can relate the amount (measured in moles) to the chemical equation.

What amount of HCl (in moles) is present in 500 cm^3 of 1-mol dm^{-3} solution?

3.17   0.5 mol, i.e. 1000 cm^3 contains 1 mol, so 500 cm^3 contains 0.5 mol.

A solution contains 40 g dm^{-3} of KCl (RMM = 74.5). What amount of KCl is present in 300 cm^3 of solution?

Hint: work as follows:
  a. What amount is present in 1000 cm^3?
  b. Hence what amount is present in 300 cm^3?

3.18   a. $\dfrac{40}{74.5}$ mol.

b. $\dfrac{40}{74.5} \times \dfrac{300}{1000} = 0.161$ mol.

Having seen these conversions, we shall put them into practice in a problem by working through it stage by stage.

Problem 3.2

0.300 g of a chromium compound was oxidized to convert the chromium to dichromate $(Cr_2O_7^{2-})$. An excess of iodide was added to the acidified solution, and the liberated iodine reacted with 27.00 cm^3 of 0.1-mol dm^{-3} thiosulphate $(S_2O_3^{2-})$. Calculate the % Cr in the compound (Cr = 52).

$$Cr_2O_7^{2-} + 6I^- + 14H^+ \rightarrow 2Cr^{3+} + 7H_2O + 3I_2 \qquad (3.3)$$

$$I_2 + 2S_2O_3^{2-} \rightarrow 2I^- + S_4O_6^{2-} \qquad (3.4)$$

Equations 3.3 and 3.4 could be worked out using the techniques covered in Programme 2, but in this programme equations will be given. These two equations, however, are not acceptable as they stand, because 3 mol of I_2 are produced in reaction 3.3 but only 1 mol of I_2 is consumed by reaction 3.4.

Make the two equations compatible with each other by making the amount of iodine the same in each case.

3.19   Multiply the second equation by 3:

$$Cr_2O_7^{2-} + 6I^- + 14H^+ \rightarrow 2Cr^{3+} + 7H_2O + 3I_2 \qquad (3.5)$$

$$3I_2 + 6S_2O_3^{2-} \rightarrow 6I^- + 3S_4O_6^{2-} \qquad (3.6)$$

The equations now tell us that 6 mol of $S_2O_3^{2-}$ react completely with the iodine liberated by 1 mol of $Cr_2O_7^{2-}$. We are therefore ready to begin the calculation. We shall need to find the following:

        a. Amount of $S_2O_3^{2-}$ (moles)⎫ These two are related

Hence   b. Amount of $Cr_2O_7^{2-}$ (moles)⎭ by the equations

Hence   c. Amount of Cr (moles)

Hence   d. Mass of Cr

Hence   e. % Cr

Work out the amount of $S_2O_3^{2-}$ used in the titration (27.00 cm³ of 0.1-mol dm⁻³ $S_2O_3^{2-}$).

3.20 $\dfrac{27.00}{1000} \times 0.1 \text{ mol} = 0.0027 \text{ mol}.$

Look back at equations 3.5 and 3.6 and work out the amount of $Cr_2O_7^{2-}$ corresponding to this. (Ask yourself whether there is more or less $Cr_2O_7^{2-}$ and by what factor).

3.21 $0.0027/6 = 0.00045$ mol of $Cr_2O_7^{2-}$ because 1 mol of $Cr_2O_7^{2-}$ produces enough iodine to react with 6 mol of $S_2O_3^{2-}$.

What amount of Cr is present in 0.00045 mol of $Cr_2O_7^{2-}$?

3.22 $2 \times 0.0045 = 0.0009$ mol, since each $Cr_2O_7^{2-}$ contains two Cr.

What is the mass of 0.0009 mol of Cr (RAM = 52)?

3.23 $0.0009 \times 52 = 0.0468$ g.

Hence, what is the % Cr in the original compound?

3.24   % $Cr = \dfrac{0.0468}{0.300} \times 100 = 15.6\%$.

This problem illustrates the basic method of tackling any numerical analytical question. Balanced chemical equations give the relationship between the numbers of reacting ions or molecules. This is the same as the relationship between the amounts measured in moles. The relative molecular masses then give mass relationships. Look back at the problem and at the outline in Frame 3.19 to see the overall tactics of solving the problem.

You should, by now, realise that there was one essential prerequisite to solving this problem, and that was the set of balanced equations. This is always vitally important in analytical problems so **BEFORE TACKLING AN ANALYTICAL PROBLEM, ALWAYS CONSTRUCT BALANCED EQUATIONS FOR ALL REACTIONS.**

The next problem is similar to the last one, so you should be able to work your way completely through it with relatively little guidance. Try to tackle it on your own before looking at the help given in the second half of Frame 3.25.

Problem 3.3

The chromium in a 1.200 g sample of a compound was converted into dichromate and the solution made up to 250 cm³. 25 cm³ of 0.1-mol dm⁻³ Fe^{2+} required 27.40 cm³ of the dichromate solution for complete oxidation to Fe^{3+}.
Calculate % Cr in the compound (Cr = 52).

$$Cr_2O_7^{2-} + 14H^+ + 6e^- \rightarrow 2Cr^{3+} + 7H_2O \qquad (3.7)$$
$$Fe^{2+} \rightarrow Fe^{3+} + e^- \qquad (3.8)$$

Just to emphasize the importance of balanced equations, combine equations 3.7 and 3.8 into one to represent the complete reaction.

3.25 $Cr_2O_7^{2-} + 14H^+ + 6e^- \rightarrow 2Cr^{3+} + 7H_2O$
$6Fe^{2+} \rightarrow 6Fe^{3+} + 6e^-$

$Cr_2O_7^{2-} + 6Fe^{2+} + 14H^+ \rightarrow 2Cr^{3+} + 6Fe^{3+} + 7H_2O$ (3.9)

The tactics of tackling this problem are similar to the previous one. We want to find the mass of Cr in the compound (and hence % Cr). We know the amount of Fe^{2+} in the titration. Equation 3.9 relates the amount of Fe^{2+} to the amount of $Cr_2O_7^{2-}$ and the RAM of Cr relates this to the mass of Cr.

Proceed as follows to solve the problem:
  a. Find the amount of Fe^{2+} in the titration (moles).
  b. Hence find the amount of $Cr_2O_7^{2-}$ in the titration.
  c. Double this to give the amount of Cr in the 27.40 cm³.
  d. Scale this up to give the amount of Cr in 250 cm³.
  e. Convert to mass and find % Cr in the compound.

3.26   The answers to the successive stages are as follows:

Amount of Fe^{2+} in titration $= \dfrac{25}{1000} \times 0.1$ mol.

Amount of $Cr_2O_7^{2-}$ in titration $= \dfrac{25}{1000} \times 0.1 \times \dfrac{1}{6}$ mol.

Amount of Cr in 27.40 cm³ $= \dfrac{25}{1000} \times 0.1 \times \dfrac{2}{6}$ mol.

Amount of Cr in 250 cm³ $= \dfrac{25}{1000} \times 0.1 \times \dfrac{2}{6} \times \dfrac{250}{27.40}$ mol.

% Cr in the compound $= \dfrac{25}{1000} \times 0.1 \times \dfrac{2}{6} \times \dfrac{250}{27.40} \times 52 \times \dfrac{100}{1.200}$

                       $= 32.95\%$.

This is a convenient place to break your study of this programme.

3.27   We shall now look at the sort of problem which involves working in the other direction. Starting from data on known compounds, we have to deduce an equation. Here we can usually use the analytical data to find the relative amounts of the reactants, and the reaction products can be found by using oxidation numbers. Suppose in this sort of example we found that $2MnO_4^-$ reacted with $5Tl^+$ in acidic solution.

How many electrons are involved in the reduction of $2MnO_4^-$ to $2Mn^{2+}$?

3.28   Ten. The oxidation state change is 5, i.e. from Mn(VII) to Mn(II), and there are 2 mol of Mn.

Hence ten electrons must be removed in the oxidation of $5Tl^+$, so what must be the final oxidation state of thallium?

3.29 $+3$, i.e.

$$5Tl^+ \rightarrow 5Tl^{3+} + 10e^-$$

We have therefore obtained the equation from the analytical data and have managed to show that the higher oxidation state of thallium is $+3$:

$$2MnO_4^- + 5Tl^+ + 16H^+ \rightarrow 2Mn^{2+} + 5Tl^{3+} + 8H_2O$$

The next problem is one of this type:

Problem 3.4

The sulphur dioxide liberated from 0.235 g of sodium disulphate(IV), $Na_2S_2O_5$, was titrated with 0.04-mol dm^{-3} $KMnO_4$ in a weakly alkaline solution. 41.30 cm^3 was required.
Find the final oxidation state of the manganese and the equation for the titration ($Na_2S_2O_5 = 190$).

$$Na_2S_2O_5 + 2H^+ \rightarrow 2Na^+ + H_2O + 2SO_2 \qquad (3.10)$$

SO_2 is oxidized to SO_4^{2-}.

We can find the equation of the reaction if we can obtain the amounts in moles of the two reactants, SO_2 and MnO_4^-.

Use the given mass of sodium disulphate(IV) together with equation 3.10 to find the amount of SO_2 involved.

3.30   Amount of $Na_2S_2O_5 = 0.235/190 = 0.00124$ mol.
        Amount of $SO_2 = 0.00124 \times 2 = 0.00248$ mol.
        Equation 3.10 indicates that one $Na_2S_2O_5$ produces two SO_2.

        Find the amount of MnO_4^- in the titration reaction.

3.31   Amount of $MnO_4^- = \dfrac{41.30}{1000} \times 0.04 = 0.00165$ mol.

        A glance at the amounts of SO_2 and MnO_4^- should indicate a reacting ratio of $3SO_2:2MnO_4^-$.

        We are told that $S^{IV}O_2$ is oxidized to $S^{VI}O_4^{2-}$. How many electrons are involved in the oxidation of $3SO_2$ to $3SO_4^{2-}$?

3.32   Six. 3 mol of SO_2 are oxidized through two oxidation states ($+4$ to $+6$).

        Hence, if 6 electrons are involved in the oxidation of $3SO_2$, 6 electrons must also be involved in the reduction of $2Mn^{VII}O_4^{2-}$.

        What is the final oxidation state of Mn?

3.33 $+4$, i.e.

$$2Mn(VII) + 6e^- \rightarrow 2Mn(IV)$$

        The most common compound of Mn(IV) is MnO_2, so we can write as our 'skelton equation'

$$3SO_2 + 2MnO_4^- \rightarrow 3SO_4^{2-} + 2MnO_2$$

        Remember that the reaction was carried out in dilute alkaline solution, and add OH^- and H_2O to obtain a balanced equation.

3.34 $$3SO_2 + 2MnO_4^- + 4OH^- \rightarrow 3SO_4^{2-} + 2MnO_2 + 2H_2O$$

(an equation involving SO_3^{2-} rather than SO_2 would be even better).

The final type of problem we shall consider is one involving a back-titration. These often cause trouble and it is frequently a good idea to work out some sort of diagrammatic scheme to illustrate what is happening.

Problem 3.5

1.000 g of a chromium compound was treated to convert the chromium into dichromate. 75 cm³ of 0.1-mol dm⁻³ Fe²⁺ were added to the acidified solution and the excess of iron needed a back-titration of 3.60 cm³ of 0.05-mol dm⁻³ $K_2Cr_2O_7$.

Calculate % Cr in the compound (Cr = 52).

$$Cr_2O_7^{2-} + 6Fe^{2+} + 14H^+ \rightarrow 2Cr^{3+} + 6Fe^{3+} + 7H_2O \quad (3.11)$$

Bearing in mind that 1 mol of $Cr_2O_7^{2-}$ contains 2 mol of Cr, equation 3.11 shows that 1 mol of Cr(VI) reacts with 3 mol of Fe(II). We can say:

$$\begin{bmatrix} \text{Cr from 1.00 g of compound} \\ \text{plus Cr from 3.60 cm}^3 \text{ of} \\ \text{0.05-mol dm}^{-3} \text{ K}_2\text{Cr}_2\text{O}_7 \end{bmatrix} \begin{array}{c} \text{reacts} \\ \text{with} \end{array} \begin{bmatrix} \text{Fe from 75 cm}^3 \text{ of} \\ \text{0.1-mol dm}^{-3} \text{ Fe}^{2+} \\ \text{solution} \end{bmatrix}$$

Calculate the amount of Fe in 75 cm³ of 0.1-mol dm⁻³ Fe²⁺ solution.

3.35 $$0.1 \times 75/1000 = 0.0075 \text{ mol.}$$

What amount of Cr(VI) will react with 0.0075 mol of Fe²⁺?

3.36 $0.0075/3 = 0.0025$ mol, since 1 mol of Cr(VI) reacts with 3 mol of Fe²⁺.

What is the amount of Cr in 3.60 cm³ of 0.05 mol dm⁻³ $K_2Cr_2O_7$? (*N.B.* Take care!).

3.37 $0.05 \times \dfrac{3.6}{1000} \times 2 = 0.00036$ (there are 2 mol of Cr in 1 mol of $K_2Cr_2O_7$).

Let us now re-write the diagram in Frame 3.34, adding the quantities calculated so far:

$$(0.0025 \text{ mol}) \left\{ \begin{array}{c} \text{Cr from 1.000 g of compound} \\ \text{plus} \\ \text{Cr from 3.60 cm}^3 \text{ of 0.05-mol} \\ \text{dm}^{-3} \text{ K}_2\text{Cr}_2\text{O}_7 \text{ (0.00036 mol)} \end{array} \right] \begin{array}{c} \text{reacts} \\ \text{with} \end{array} \left[\begin{array}{c} \text{Fe from 75 cm}^3 \text{ of} \\ \text{0.1-mol dm}^{-3} \text{ Fe}^{2+} \\ \text{solution (0.0075 mol)} \end{array} \right]$$

What is the amount of Cr in 1.000 g of the unknown compound?

3.38 $0.0025 - 0.00036 \text{ mol} = 0.00214 \text{ mol}$.

What is the % Cr in the compound?

3.39   11.13%, i.e.
Mass of Cr $= 0.00214 \times 52$.
$$\% \text{ Cr} = 0.00214 \times 52 \times \frac{100}{1.000} = 11.13\%.$$

Analytical Calculations Test

Numbers correspond to the programme objectives.
1. What amount of KNO_3 is there in
   a. 75 g of solid potassium nitrate?
   b. 25 cm^3 of 0.02-mol dm^{-3} KNO_3 solution
      ($KNO_3 = 101$)?

2. What mass of sulphur is there in
   a. 27 g of $K_2S_2O_8$?
   b. 50 cm^3 of 0.1-mol dm^{-3} $Na_2S_2O_3$
      ($K_2S_2O_8 = 270$, $Na_2S_2O_3 = 158$, $S = 32$)?
3. A compound was analysed gravimetrically for sulphate by converting all the sulphate into $BaSO_4$. 0.220 g of the compound gave 0.3610 g of $BaSO_4$. Calculate % SO_4^{2-} in the compound ($Ba = 137.3$, $S = 32$, $O = 16$).
4. a. A sample of 0.100 g of a hydrazine salt was titrated with potassium iodate solution containing 5.200 g dm^{-3} of KIO_3 under conditions which gave I^+ (as ICl_2^-) as the final iodine product. This required 31.60 cm^3 of iodate(V) solution.
      Calculate % N_2H_4 in the salt. ($N = 14$, $H = 1$, $K = 39$, $I = 127$, $O = 16$).

   Method: Find the number of electrons involved in oxidizing N_2H_4 and in reducing IO_3^-. Hence obtain a balanced equation. Find the amount of IO_3^- in the titration and use the chemical equation to find the amount of N_2H_4 in the salt. Hence calculate %N_2H_4.
   b. Salts of thallium(I) are oxidized by acidic KIO_3, which is reduced to the iodide. 25 cm^3 of a solution containing 12.72 g dm^{-3} of Tl_2SO_4 required 25.25 cm^3 of 0.01667-mol dm^{-3} KIO_3. Find the equation for the reaction between Tl^+ and KIO_3 ($Tl = 204.4$, $S = 32$, $O = 16$).
5. 25 cm^3 of a solution containing 2 g dm^{-3} of hydroxylamine (NH_2OH) was boiled with an excess of iron(III) sulphate in acidic solution. The resulting solution contained iron(II) equivalent to 30.3 cm^3 of 0.02-mol dm^{-3} $KMnO_4$. Obtain an equation for the reaction between iron(III) sulphate and hydroxylamine.

$$MnO_4^- + 5Fe^{2+} + 8H^+ \rightarrow Mn^{2+} + 5Fe^{3+} + 4H_2O$$

($N = 14$, $H = 1$, $O = 16$).

Answers

1. a. $75/101 = 0.742$ mol.

    1

   b. $0.02 \times \dfrac{25}{1000} = 0.0005$ mol.

    1

2. a. $27 \times \dfrac{64}{270} = 6.4$ g.

    1

   b. $64 \times 0.1 \times \dfrac{50}{1000} = 0.32$ g.

    1

3. Mass of $SO_4^{2-} = 0.3610 \times \dfrac{96 \text{ g}}{233.3}$.

    2

   % of $SO_4^{2-} = 0.3610 \times \dfrac{96}{233.3} \times \dfrac{100}{0.220} = 67.5\%$

4. a. Oxidation: $N_2H_4 \rightarrow N_2 + 4H^+ + 4e^-$
    Reduction: $IO_3^- + 4e^- + 6H^+ \rightarrow 3H_2O + I^+$
    Hence the balanced equation is:
    $IO_3^- + N_2H_4 + 2H^+ \rightarrow I^+ + N_2 + 3H_2O$

    2

   Amount of IO_3^- in 1 dm^3 $= \dfrac{5.200}{214}$ mol

   Amount of IO_3^- in titration $= \dfrac{5.200}{214} \times \dfrac{31.6}{1000}$ mol

   $\therefore$ Amount of N_2H_4 in 0.1 g of salt $= \dfrac{5.200}{214} \times \dfrac{31.6}{1000}$ mol

    2

   Mass of $N_2H_4 = \dfrac{5.200}{214} \times \dfrac{31.6}{1000} \times 32$ g

    1

   $\% N_2H_4 = \dfrac{5.200}{214} \times \dfrac{31.6}{1000} \times 32 \times \dfrac{100}{0.1} = 24.6\%$

    2

   b. Amount of $KIO_3 = 0.01667 \times \dfrac{25.25}{1000} = 0.000421$ mol

    2

   Amount of Tl^+ in 25 cm^3 of solution

        $= \dfrac{12.72}{504.8} \times 2 \times \dfrac{25}{1000} = 0.001260$ mol

    2

   Reacting mole ratio $= KIO_3 : 3Tl^+$

    1

   Oxidation state change for $I = 6$

   $\therefore$ Oxidation state change for $Tl = 2$

    2

   $3Tl^+ + KIO_3 + 6H^+ \rightarrow 3Tl^{3+} + KI + 3H_2O$

    2

5. Amount of MnO_4^- in titration $= 0.02 \times \dfrac{30.3}{1000}$ mol

Amount of Fe^{2+} produced by reduction

$$= 0.02 \times \frac{30.3}{1000} \times 5 = 0.00303 \text{ mol} \qquad 2$$

Amount of NH_2OH in reaction $= \dfrac{2}{33} \times \dfrac{25}{1000} = 0.001515$ mol   2

Reacting mole ratio $= NH_2OH : 2Fe^{3+}$ 1
Oxidation state change of Fe $= 1$ 1
Since there are 2 mol of Fe for each N,
oxidation state change of N $= 2$ 1

$$\overset{-1}{NH_2OH} \rightarrow \overset{+1}{N_2O}$$

$2Fe_2(SO_4)_3 + 2NH_2OH \rightarrow 4FeSO_4 + N_2O + 2H_2SO_4 + H_2O$ 2

Total   33

The other two programmes in this set do not develop analytical calculations any further, so this test is your own measure of how well you have understood the topic. Questions 3, 4 and 5 are clearly the most important and correct solutions to these (particularly 4b and 5, which are fairly difficult) indicate a good understanding of the topic.

Analytical Calculations

Revision Notes

Analytical calculations are basically simple, provided a few fundamentals are adhered to:
   i. Always construct a balanced equation.
   ii. Always work in moles, not grams.
   iii. Plan your way through the problem at the start.

There are a number of rather self-evident relationships between the quantities involved in analytical calculations:

$$\text{Amount/mol} = \frac{\text{mass/g}}{\text{Relative molecular mass/g mol}^{-1}}$$

$$\text{Concentration/mol dm}^{-3} = \text{amount/mol} \times \frac{1000}{\text{volume/cm}^3}$$

It is always essential to refer to a formula and not a name when talking about amounts in moles, e.g. 1 mol of H weighs 1 g, but 1 mol of H_2 weighs 2 g. 'One mole of hydrogen' is meaningless. Always remember such points as 1 mol of $K_2Cr_2O_7$ contains 2 mol of Cr.

Programme 4

Thermodynamic Aspects

Objectives

After completing this programme, you should be able to:
1. Explain the meaning of electrode potential.
2. Explain why electropositive metals have negative electrode potentials.
3. Use electrode potential data to predict the likely occurrence of redox reactions.
4. Calculate ΔG and K from standard electrode potential data.
5. Calculate the standard electrode potential of a half reaction, given the standard electrode potential of two related half reactions.
6. State the effect of added reagents on electrode potential, and calculate electrode potential under non-standard conditions.

Objectives 3–6 are tested at the end of the programme.

Assumed Knowledge

Before starting the programme you should be able to balance oxidation–reduction equations and write ion-electron half equations for oxidation and reduction processes. You should also recognize and be able to use the expression relating equilibrium constant to standard Gibbs free energy change:

$$\Delta G^{\ominus} = -RT \ln K$$

Thermodynamic Aspects

4.1   In programme 1 we met three ideas basic to the subject of oxidation and reduction, namely:
  i.   Oxidation and reduction always occur together.
  ii. Oxidizing agents can be put in order of oxidizing power.
  iii. Oxidation and reduction can be regarded as electronic processes.

In the following equation, state what is being oxidized and what is being reduced:

$$3Cu + 8HNO_3 \rightarrow 3Cu(NO_3)_2 + 2NO\uparrow + 4H_2O \qquad (4.1)$$

4.2   Cu is oxidized to Cu^{2+}.
Some of the HNO_3 is reduced to NO.

Look at the following equation:

$$4H_2O + 3Fe \rightarrow Fe_3O_4 + 4H_2 \qquad (4.2)$$

Which of the two oxides in this equation is the more powerful oxidant?

4.3 H_2O, because it oxidizes iron to Fe_3O_4.

Define oxidation and reduction in electronic terms.

4.4   Oxidation is electron loss.
Reduction is electron gain.

If you have correctly answered these questions, you can proceed with this programme, otherwise you should return to Programme 1—Basic Principles.

Since oxidation and reduction can be regarded as electronic processes, it should be possible to measure oxidizing and reducing power by using electrical methods. We might, for instance, be able to measure an electrical potential which will give an indication of how strong an oxidant a compound is.

Let us take zinc as an example. Zinc metal undergoes many reactions in which it becomes zinc ions in solution. Write the half-equation for this process and say whether the zinc is acting as an oxidizing agent or a reducing agent.

4.5 $$Zn \rightarrow Zn^{2+}(aq) + 2e^- \qquad (4.3)$$

Zinc is acting as a reducing agent, since it is giving electrons to something else and reducing it. Alternatively, we can say that the zinc is being oxidized from the zero to the $+2$ oxidation state, so something else must be reduced at the same time.

Let us consider, now, what can be best termed a 'thought experiment'. If we dip a piece of zinc into water, zinc ions will tend to go into solution, leaving the electrons on the metal. Draw a diagram to show this state of affairs.

4.6   Zn   We now have a solution containing ions and a strip of metal containing excess electrons. Is the electric potential of the metal positive or negative relative to the solution?

4.7 Negative, since the metal carries excess electrons. We call this potential the electrode potential of zinc under these conditions. Note particularly that zinc is an electropositive metal, but has a negative electrode potential. This is because the term electropositive means it has a relatively high tendency to go into solution as the positive ion, and so leaves behind a relatively large excess of negative electrons on the electrode.

So far we are still in the realms of 'thought chemistry', but if we tried to measure the electrode potential of zinc with a suitable meter we would have to make a connection to the solution with a metal electrode. This electrode, possibly made of copper, would have its own electrode potential.

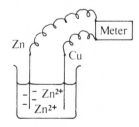

This experiment, therefore, would not actually measure the electrode potential of zinc. What would it measure?

4.8   Since the copper has its own electrode potential, this experiment would measure the *difference* in electrode potential between zinc and copper.

This is clearly a problem that we cannot get over. Any experiment will always measure the difference between two electrode potentials. We must therefore arbitrarily define one potential to be zero. The one chosen is fairly convenient for most aqueous solution work, and is

$$H^+(aq) \text{ (unit activity)} + e^- \rightarrow \tfrac{1}{2}H_2(g) \text{ (unit activity)} \quad (4.4)$$

The potential of this half reaction is defined to be zero, and all other electrode potentials are referred to it. (The actual construction of a standard hydrogen electrode will be found in most standard texts; see, e.g. A. G. Sharpe, *Principles of Oxidation and Reduction,* RIC Monographs for Teachers, No.2.)

Which of the species H^+ and H_2 is, in principle, an oxidizing agent, and which is a reducing agent?

4.9 H^+ is the oxidizing agent.
H_2 is the reducing agent.

Write the equation for the oxidation of zinc by H^+.

4.10 $$Zn + 2H^+ \rightarrow Zn^{2+} + H_2 \quad (4.5)$$

Since the electrode potential of zinc is lower than that of hydrogen, hydrogen in its oxidized form (H^+) will oxidize zinc in its reduced form (Zn). The electrode potential of copper, on the other hand, is above that of hydrogen, so the reverse reaction should, in principle, be possible. The reaction between H_2 and Cu^{2+} is, however, very slow, although a very similar reaction can be performed:

$$H_2 + CuO \rightarrow H_2O + Cu \quad (4.6)$$

Write the equivalent reaction for calcium and state the direction in which it will go.

4.11

or

$$Ca + H_2O \rightarrow H_2 + CaO \qquad (4.7)$$

$$Ca + 2H_2O \rightarrow H_2 + Ca(OH)_2 \qquad (4.8)$$

i.e. opposite to the copper reaction.

The electrode potential of hydrogen is defined to be zero, that of copper is positive; what is the probable sign of the electrode potential of calcium?

4.12   Negative, i.e. calcium in its reduced form (Ca) reduces hydrogen from its oxidized (H_2O) to its reduced (H_2) form.

The actual values of the Standard Electrode Potentials, i.e. when all substances are at unit activity, are as follows:

$$Ca^{2+}(aq) + 2e^- \rightarrow Ca(s) \qquad E^\ominus = -2.87 \text{ V}$$

$$Cu^{2+}(aq) + 2e^- \rightarrow Cu(s) \qquad E^\ominus = +0.34 \text{ V}$$

$$H^+(aq) + e^- \rightarrow \tfrac{1}{2}H_2(g) \qquad E^\ominus = 0 \text{ V}$$

Let us now turn to the oxidizing power of substances such as manganate(VII) and chromate(VI). Here there is no free metal involved in the half reaction.

Write the half equation for the reduction manganate(VII) to Mn^{2+} in acidic solution.

4.13

$$MnO_4^- + 8H^+ + 5e^- \rightarrow Mn^{2+} + 4H_2O \qquad (4.9)$$

To see how we can measure the oxidizing power of manganate(VII) in terms of an electrical potential we must resort to another 'thought experiment'. If we dip an inert electrode (such as Pt) into a solution of manganate(VII), the above half reaction tends to proceed to a very small extent and the electrons required come from the platinum.

What, therefore, is the sign of the electrical potential of the platinum at this stage?

4.14   Withdrawal of electrons from the Pt leaves it with a positive potential. Electrons are withdrawn until the potential reaches such a level (the electrode potential) that the process is stopped.

Hence our electrical method of assessing oxidizing power can also be extended to systems where a metallic element is not a reactant or product. Again a positive electrode potential indicates a stronger oxidant than the reference hydrogen electrode.

Electrode potentials vary as the concentration of reactants varies. Consider equation 4.9 as an equilibrium:

$$MnO_4^- + 8H^+ + 5e^- \rightleftharpoons Mn^{2+} + 4H_2O \qquad (4.9)$$

Clearly, if the concentration of MnO_4^- is high, the equilibrium will be displaced to the right, more electrons will be removed from the inert platinum electrode and the electrode potential will therefore be more positive. The values quoted in tables are therefore **STANDARD ELECTRODE POTENTIALS**, i.e. the potential of the electrode relative to the standard hydrogen electrode when all reactants and products are at unit activity. Standard electrode potentials are given the symbol $E^{\ominus}$ and are commonly quoted as follows:

Zn^{2+}/Zn, $E^{\ominus} = -0.76$ V

or $Zn^{2+}(aq) + 2e^- \rightarrow Zn(s)$ $E^{\ominus} = -0.76$ V

$Mn(VII)/Mn(II)$, $E^{\ominus} = +1.52$ V

or $MnO_4^- + 8H^+ + 5e^- \rightarrow Mn^{2+} + 4H_2O$ $E^{\ominus} = +1.52$ V

Note that the potential is always written for the change from the high to the low oxidation state, i.e. with the electron(s) on the left of the equation.

Some standard electrode potentials are given in the Appendix to this programme. The order of magnitude of standard electrode potential in aqueous solution is from about $+2$ to -0.7 V. Any compound outside these limits is either so strong an oxidant that it readily liberates oxygen from water, or so strong a reducing agent that it liberates hydrogen.

Which of the following systems do you feel, intuitively, will have the larger electrode potential, or will they be the same:

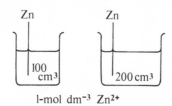

Zn   Zn

100 cm³   200 cm³

1-mol dm⁻³ Zn²⁺

4.15   They will both be the same. The potential generated when zinc goes
        into solution as zinc ions does not depend on how much zinc solution
        is present, because the process occurs to such a minute extent.

        Electrode potential is therefore like the properties in one of the follow-
        ing lists. Which?

List A	List B
Mass	Density
Length	Temperature
Volume	Heat capacity
Energy	Pressure

4.16   List B.

        All the properties in list B are termed **INTENSIVE PROPERTIES**,
        while those in list A are termed **EXTENSIVE PROPERTIES**. Exten-
        sive properties, such as mass, depend on the size or extent of the
        system, and we can simply add them up as necessary. For example,
        two masses of 50 and 30 g give a combined mass of 80 g. Intensive
        properties (list B) cannot be added up. (Consider the result of mixing
        two beakers of water at 50 and 30 °C). Electrode potentials, as inten-
        sive properties, cannot be combined by simple addition. If, therefore,
        we want to compare oxidizing or reducing power, calculate
        equilibrium constants for redox reactions, etc., we must convert the
        intensive property standard electrode potential to a suitable extensive
        property.

        Fortunately, standard electrode potential is simply related to standard
        free energy change, which is an extensive property from which we can
        calculate equilibrium constants. The relationship will not be proved
        here, but is shown in many standard texts to be

$$\Delta G^{\ominus} = -nFE^{\ominus}$$

        where

        n = number of electrons in the half equation;
        F = Faraday constant (96 487 C mol^{-1});
        $E^{\ominus}$ = standard electrode potential in volts (JC^{-1}).

        Use the above relationship to calculate $\Delta G^{\ominus}$ for the half reaction.

$$Fe^{2+}(aq) + 2e^{-} \rightarrow Fe(s) \quad E^{\ominus} = -0.44 \text{ V} \quad (4.10)$$

4.17 85 kJ mol^{-1}, i.e.

$$\Delta G^{\ominus} = -2 \times 96487 \text{ C mol}^{-1} \times (-0.44) \text{ J C}^{-1}$$
$$= +2 \times 96487 \times 0.44 \text{ J mol}^{-1}$$
$$= 85 \text{ kJ mol}^{-1}$$

It must be remembered that since $E^{\ominus}$ is relative to the standard hydrogen electrode, the resulting $\Delta G^{\ominus}$ is also relative to the hydrogen half-reaction, i.e.

$$Fe^{2+}(aq) + 2e^- \rightarrow Fe(s)$$
$$H_2 \rightarrow 2H^+(aq) + 2e^-$$

$$Fe^{2+}(aq) + H_2 \rightarrow Fe(s) + 2H^+(aq) \tag{4.11}$$

Strictly, $\Delta G^{\ominus}$ for equation 4.11 is $+85 \text{ kJ mol}^{-1}$, but this value is often loosely applied to equation 4.10.

The two items of data ($E^{\ominus}$ and $\Delta G^{\ominus}$) for this reaction both convey similar information. $E^{\ominus}$ is negative, which can be loosely interpreted to mean that the reaction does not tend to proceed in the direction written, and the same conclusion can be drawn from the positive standard free energy.

N.B. When using the relationship $\Delta G^{\ominus} = -nFE^{\ominus}$, it must be remembered that $E^{\ominus}$ is always quoted for the half reaction with the electrons on the left of the equation. The sign of $\Delta G^{\ominus}$ must clearly be reversed if the direction of reaction is reversed.

Problem 4.1

Calculate $\Delta G^{\ominus}$ for

$$Tl^{3+} + 2I^- \rightarrow Tl^+ + I_2$$

given
$$Tl^{3+} + 2e^- \rightarrow Tl^+ \qquad\qquad E^{\ominus} = +1.26 \text{ V}$$
$$\tfrac{1}{2}I_2 + e^- \rightarrow I^- \qquad\qquad E^{\ominus} = +0.54 \text{ V}$$

Start this problem by splitting the complete equation into two ion-electron half equations.

4.18
$$Tl^{3+} + 2e^- \rightarrow Tl^+ \qquad (4.12)$$
$$2I^- \rightarrow I_2 + 2e \qquad (4.13)$$

$$Tl^{3+} + 2I^- \rightarrow Tl^+ + I_2 \qquad (4.14)$$

Use the relationship $\Delta G^{\ominus} = -nFE^{\ominus}$ (Frame 4.16) to find $\Delta G^{\ominus}$ for equation 4.12.

4.19 $\Delta G^{\ominus} = -2 \times 96487 \times 1.26 \text{ J mol}^{-1} = -243 \text{ kJ mol}^{-1}$.

Similarly, find $\Delta G^{\ominus}$ for

$$\tfrac{1}{2}I_2 + e^- \rightarrow I^- \qquad E^{\ominus} = +0.54 \text{ V} \qquad (4.15)$$

4.20 $\Delta G^{\ominus} = -1 \times 96487 \times 0.54 \text{ J mol}^{-1} = -52 \text{ kJ mol}^{-1}$.

Equation 4.13 is obtained by reversing and doubling equation 4.15, so what is the standard free energy change for equation 4.13?

4.21 $+104 \text{ kJ mol}^{-1}$, i.e. the standard free energy change is also doubled and its sign reversed.

Equation 4.14 is the sum of equations 4.12 and 4.13, so find the value of $\Delta G^{\ominus}$ for equation 4.14.

4.22

$$Tl^{3+} + 2e^- \rightarrow Tl^+ \qquad \Delta G^\ominus = -243$$
$$2I^- \rightarrow I_2 + 2e^- \qquad \Delta G^\ominus = +104$$

$$Tl^{3+} + 2I^- \rightarrow Tl^+ + I_2 \qquad \Delta G^\ominus = -139 \text{ kJ mol}^{-1}$$

The large negative standard free energy change suggests that the reaction, at equilibrium, will have proceeded virtually completely to Tl^+ and I_2. The free energy change tells us nothing, of course, about how quickly equilibrium will be reached, only about the ultimate position of the equilibrium. The same conclusion could have been drawn (less precisely) from the $E^\ominus$ data, since if two standard potentials differ by about 0.4 V or more, the associated redox reaction goes essentially to completion.

Problem 4.2

Calculate $E^\ominus$ for:
$$Fe^{3+}(aq) + 3e^- \rightarrow Fe(s) \qquad (4.16)$$
given $\quad Fe^{3+}(aq) + e^- \rightarrow Fe^{2+}(aq) \qquad E^\ominus = +0.76 \text{ V} \qquad (4.17)$
$\qquad\quad Fe^{2+}(aq) + 2e^- \rightarrow Fe(s) \qquad E^\ominus = -0.44 \text{ V} \qquad (4.18)$

This is the type of problem which immediately invites simple addition of $E^\ominus$ values. The correct method of tackling it (via $\Delta G^\ominus$) gives a different answer, which is in fact the right answer.

Clearly, equation 4.16 is simply the sum of equations 4.17 and 4.18, so find $\Delta G^\ominus$ for equations 4.17 and 4.18 and add them to get $\Delta G^\ominus$ for equation 4.16. (The numerical value of F need not be used.)

4.23 $\Delta G^\ominus = (-1 \times F \times 0.76) + (-2 \times F \times -0.44) \text{ J mol}^{-1}$
$\qquad = -0.76F + 0.88F = +0.12F \text{ J mol}^{-1}$

Put this equal to $-nFE^\ominus$ to obtain $E^\ominus$ for equation 4.16.

$$0.12F = -3 \times F \times E^\ominus$$
$$E^\ominus = -0.04 \text{ V}$$
$$\text{for} \quad Fe^{3+}(aq) + 3e^- \rightarrow Fe(s) \quad (4.16)$$

Clearly, simple addition of the $E^\ominus$ values would not have given the right answer because the free energy change is dependent on the number of electrons in the equation.

We have, so far, considered only standard electrode potentials and standard free energy changes, i.e. under conditions where all reactants and products are at unit activity. We shall turn now to the more realistic state of affairs of non-standard conditions. Consider the half equation:

$$MnO_4^- + 8H^+ + 5e^- \rightarrow Mn^{2+} + 4H_2O \quad E^\ominus = +1.52 \text{ V} \quad (4.19)$$

If we increase the concentration of the MnO_4^- ion, the reaction will tend to proceed more to the right, more electrons will tend to be removed from any electrode dipped into the solution and the electrode will thus be raised to a more positive potential. The reverse effect will occur if reaction product (Mn^{2+}) is added.

This is expressed in the Nernst equation:

$$E = E^\ominus + \frac{RT}{nF} \ln\left(\frac{[\text{oxidized species}]}{[\text{reduced species}]}\right)$$

or

$$E = E^\ominus + \frac{RT}{5F} \ln\left(\frac{[MnO_4^-][H^+]^8}{[Mn^{2+}]}\right) \quad (4.20)$$

where R is the gas constant, T the temperature (K), F the Faraday constant and n the number of electrons in the half equation. The square brackets represent the activities of the species shown, although concentrations are frequently used as an approximation.

The oxidizing power of manganate(VII) is clearly heavily dependent on hydrogen ion concentration.

Look back at equations 4.19 and 4.20 and write the Nernst equation for the half reaction:

$$Cr_2O_7^{2-} + 14H^+ + 6e^- \rightarrow 2Cr^{3+} + 7H_2O \quad (4.21)$$

4.25
$$E = E^{\ominus} + \frac{RT}{6F} \ln\left(\frac{[Cr_2O_7^{2-}][H^+]^{14}}{[Cr^{3+}]^2}\right)$$

The Nernst equation can be re-written as:

$$E = E^{\ominus} + \frac{2.303RT}{nF} \log\left(\frac{[\text{oxidized species}]}{[\text{reduced species}]}\right)$$

and at 25 °C:

$$\frac{2.303RT}{F} = 0.0591 \text{ V}$$

Hence, at 25 °C:

$$E = E^{\ominus} + \frac{0.0591}{n} \log\left(\frac{[\text{oxidized species}]}{[\text{reduced species}]}\right)$$

This is usually the most convenient form of the Nernst equation.

Use the Nernst equation to calculate E at 25 °C for the half reaction

$$Ce^{4+} + e^- \rightarrow Ce^{3+} \qquad E^{\ominus} = +1.45 \text{ V} \qquad (4.22)$$

if $[Ce^{4+}] = 0.1 \text{ mol dm}^{-3}$ and $[Ce^{3+}] = 0.01 \text{ mol dm}^{-3}$.

4.26 $E = 1.509$ V, i.e.

$$E = 1.45 + \frac{0.0591}{1} \log\left(\frac{[0.1]}{[0.01]}\right)$$

$$= 1.45 + \frac{0.0591}{1} \log 10 = 1.45 + 0.0591 = 1.509 \text{ V}$$

The actual free energy change is related to the actual electrode potential (under non-standard conditions) by

$$\Delta G = -nFE$$

i.e. the same relationship as between the standard quantities.

We shall now turn to the final type of problem—finding an equilibrium constant for a redox reaction. This is most useful in assessing quantitatively the extent to which a reaction will proceed. The equilibrium constant, K, for a reaction is related to the standard free energy change, $\Delta G^{\ominus}$, by the equation

$$\Delta G^{\ominus} = -RT \ln K$$

where R is the gas constant $(= 8.312 \text{ J K}^{-1}\text{mol}^{-1})$ and T is temperature (K).

We have already seen (Frames 4.17–4.22) how to calculate $\Delta G^{\ominus}$ for a redox reaction, given the standard reduction potentials for the two half reactions. In those frames, we found

$$Tl^{3+} + 2I^- \rightleftharpoons Tl^+ + I_2 \quad \Delta G^{\ominus} = -139 \text{ kJ mol}^{-1}$$

Write the expression for the equilibrium constant of this reaction in terms of the concentrations of the four species involved.

4.27 $K = \dfrac{[Tl^+][I_2]}{[Tl^{3+}][I^-]^2}$

i.e. the higher value of K, the higher are the equilibrium concentrations of the products, Tl^+ and I_2.

Use the equation $\Delta G^{\ominus} = -RT \ln K$ to find a value for K at 25 °C, given that $\Delta G^{\ominus} = -139\,000 \text{ J mol}^{-1}$, $R = 8.312 \text{ J K}^{-1}\text{mol}^{-1}$ and $T = 298$ K.

4.28 $-139\,000 = -8.312 \times 298 \times \ln K$

$$\ln K = \frac{139\,000}{8.312 \times 298} = 56.12$$

$$K = 2.35 \times 10^{24}$$

i.e. the equilibrium constant is very large and the reaction proceeds essentially completely to the right.

Use the same method to find K for the reaction in problem 4.3.

Problem 4.3

Calculate K at $25\,^{\circ}$C for the reaction:

$Ce^{4+} + Fe^{2+} \rightleftharpoons Ce^{3+} + Fe^{3+}$ (4.23)

$Ce^{4+} + e^- \rightarrow Ce^{3+}$ $E^{\ominus} = +1.45$ V   (4.24)

$Fe^{3+} + e^- \rightarrow Fe^{2+}$ $E^{\ominus} = +0.76$ V   (4.25)

$F = 96487$ C mol^{-1}

Use the following steps to solve the problem:

  i.   Rearrange the given reactions so that they add up to equation 4.23.

  ii.   Find $\Delta G^{\ominus}$ for each of the rearranged reactions.

  iii.   Add these $\Delta G^{\ominus}$ values to find the value for equation 4.23.

  iv.   Calculate K.

4.29 $K = 4.7 \times 10^{11}$, i.e.

$$Ce^{4+} + e^- \rightarrow Ce^{3+} \qquad \Delta G^\circ = -1.45\,F$$
$$Fe^{2+} \rightarrow Fe^{3+} + e^- \qquad \Delta G^\circ = +0.76\,F$$

$$\overline{Ce^{4+} + Fe^{2+} \rightleftharpoons Ce^{3+} + Fe^{3+} \qquad \Delta G^\circ = -0.69\,F}$$

$$-0.69 \times 96\,487 = -8.312 \times 298 \times \ln K$$
$$\ln K = 26.88$$
$$K = 4.7 \times 10^{11}$$

i.e. the reaction proceeds essentially to completion. There is an alternative, equally valid, method for tackling a problem of this type using the Nernst equation (Frames 4.24–4.26). A redox reaction proceeds if the potentials of the two half reactions are different. During the reactions, the concentrations of reactants and products change until a point is reached when the two potentials are equal. At this point equilibrium is reached and the reaction stops. We can find the concentrations (or at least the ratio of concentrations) corresponding to this situation by using the Nernst equation.

Apply the Nernst equation to the reduction of cerium(IV). The half reaction is given in Equation 4.24.

4.30
$$E = E^\circ + \frac{RT}{F} \ln\left(\frac{[Ce^{4+}]}{[Ce^{3+}]}\right)$$

$$= 1.45 + 0.0591 \log\left(\frac{[Ce^{4+}]}{[Ce^{3+}]}\right) \qquad (4.26)$$

Do the same thing for the oxidation of iron(II) given in equation 4.25.

4.31
$$E = E^\circ + \frac{RT}{F} \ln\left(\frac{[Fe^{3+}]}{[Fe^{2+}]}\right)$$

$$= 0.76 + 0.0591 \log\left(\frac{[Fe^{3+}]}{[Fe^{2+}]}\right) \qquad (4.27)$$

Write out the equilibrium constant expression for equation 4.23:

$$Ce^{4+} + Fe^{2+} \rightleftharpoons Ce^{3+} + Fe^{3+} \qquad (4.23)$$

4.32
$$K = \frac{[Ce^{3+}][Fe^{3+}]}{[Ce^{4+}][Fe^{2+}]} \qquad (4.28)$$

When this reaction has reached equilibrium, the E values given by equations 4.26 and 4.27 are equal. We can therefore say that, at equilibrium

$$1.45 + 0.0591 \log\left(\frac{[Ce^{4+}]}{[Ce^{3+}]}\right) = 0.76 + 0.0591 \log\left(\frac{[Fe^{3+}]}{[Fe^{2+}]}\right)$$

Rearrange this equation to obtain a value for the equilibrium constant expression in equation 4.28. [N.B. $\log A - \log B = \log (A/B)$].

4.33
$$1.45 - 0.76 = 0.0591 \log \left(\frac{[Fe^{3+}][Ce^{3+}]}{[Fe^{2+}][Ce^{4+}]}\right)$$

$$= 0.0591 \log K$$
$$K = 4.7 \times 10^{11}$$

i.e. the same as the value obtained via $\Delta G^{\ominus}$

Thermodynamic Aspects Test

$F = 96487 \ C\,mol^{-1}$; $2.303 \ RT/F = 0.0591$ V at $25\,°C$; $R = 8.314 \ J\,K^{-1}\,mol^{-1}$.

1. The standard electrode potential for Fe^{3+}/Fe^{2+} is $+0.76$ V. For each of the following systems, state whether:
   i.   the reduced form is likely to be extensively oxidized by Fe^{3+};
   ii. the oxidized form is likely to be extensively reduced by Fe^{2+}; or
   iii. there is likely to be an equilibrium involving appreciable concentrations of all species.

   A. Ag^+/Ag $E^⦵ = +0.80$ V
   B. $\frac{1}{2}Cl_2/Cl^-$ $E^⦵ = +1.36$ V
   C. $Cr_2O_7^{2-}/2Cr^{3+}$ $E^⦵ = +1.33$ V
   D. Cu^{2+}/Cu $E^⦵ = +0.34$ V
   E. Fe^{2+}/Fe $E^⦵ = -0.44$ V
   F. Zn^{2+}/Zn $E^⦵ = -0.76$ V

2. Given the standard potentials:
$$Fe^{3+} + e^- \rightarrow Fe^{2+} \qquad E^⦵ = +0.76 \text{ V}$$
$$\tfrac{1}{2}I_2 + e^- \rightarrow I^- \qquad E^⦵ = +0.54 \text{ V}$$

   Calculate $\Delta G^⦵$ and K at $25\,°C$ for the reaction:
$$FeI_3 \rightleftharpoons FeI_2 + \tfrac{1}{2}I_2$$

3. Given the standard potentials
$$Tl^+ + e^- \rightarrow Tl \qquad E^⦵ = -0.34 \text{ V}$$
$$Tl^{3+} + 3e^- \rightarrow Tl \qquad E^⦵ = +0.73 \text{ V}$$

   Calculate $E^⦵$ for:
$$Tl^{3+} + 2e^- \rightarrow Tl^+$$

4. State whether the added reagents indicated will
   i.   raise,
   ii. lower or
   iii. have no effect on the potentials given:
      A. $Cr_2O_7^{2-}/Cr^{3+}$ $E^⦵ = +1.33$ V. Addition of OH^-
      B. MnO_4^-/Mn^{2+} $E^⦵ = +1.52$ V. Addition of H^+
      C. Ce^{4+}/Ce^{3+} $E^⦵ = +1.45$ V. Addition of Zn^{2+}
      D. Fe^{3+}/Fe^{2+} $E^⦵ = +0.76$ V. Addition of Fe^{2+}

   Calculate E for $VO^{2+} + e^- + 2H^+ \rightarrow V^{3+} + H_2O$ at $25\,°C$ when $[H^+] = 0.1 \ mol\,dm^{-3}$ and nine tenths of the vanadium present has been reduced from the $+4$ to the $+3$ state. $E^⦵$ for V^{IV}/V^{III} is $+0.34$ V.

Answers

Marks

1. A. Equilibrium.   1
    B. Cl_2 extensively reduced by Fe^{2+}.   1
    C. $Cr_2O_7^{2-}$ extensively reduced by Fe^{2+}.   1
    D. Cu extensively oxidised by Fe^{3+}.   1
    E. Fe extensively oxidised by Fe^{3+}.   1
    F. Zn extensively oxidised by Fe^{3+}.   1

2. $FeI_3 \rightleftharpoons FeI_2 + \frac{1}{2}I_2$ becomes
    $Fe^{3+} + 3I^- \rightarrow Fe^{2+} + 2I^- + \frac{1}{2}I_2$

$Fe^{3+} + e^- \rightarrow Fe^{2+}$	$\Delta G^{\ominus} = -0.76F$	1
$I^- \rightarrow \frac{1}{2}I_2 + e^-$	$\Delta G^{\ominus} = +0.54F$	1
$Fe^{3+} + I^- \rightarrow Fe^{2+} + \frac{1}{2}I_2$	$\Delta G^{\ominus} = -0.22F = -21.2\text{ kJ mol}^{-1}$	1
	$K = 5.3 \times 10^3$	2

3. | | | |
    |---|---|---|
    | $Tl^{3+} + 3e^- \rightarrow Tl$ | $\Delta G^{\ominus} = -3 \times 0.73F = -2.19F$ | |
    | $Tl \rightarrow Tl^+ + e^-$ | $\Delta G^{\ominus} = -0.34F$ | |
    | $Tl^{3+} + 2e^- \rightarrow Tl^+$ | $\Delta G^{\ominus} = -2.53F$ | 2 |
    | | $E^{\ominus} = +1.26\text{ V}$ | 2 |

4. A. Lower (it consumes H^+, which is a reactant).   1
    B. Raise (opposite of A).   1
    C. No effect (Zn^{2+} is inert).   1
    D. Lower (Fe^{2+} is a reaction product).   1

$$E = 0.34 + 0.0591 \log\left(\frac{[H^+]^2[VO^{2+}]}{[V^{3+}]}\right)$$ 2

$$= 0.34 + 0.0591 \log\left(\frac{0.01 \times 0.1}{0.9}\right)$$

$$= 0.34 + 0.0591 \times (-2.954)$$

$$= 0.34 - 0.175 = 0.165\text{ V}$$ 2

Total   23

Thermodynamic aspects of oxidation and reduction are not developed further in this set of programmes, so this test is your own measure of how well you have understood the topic. All questions are important in demonstrating different aspects of the subject and an overall score of 18 or more indicates a reasonable understanding.

Thermodynamic Aspects

Revision Notes

A metal dipped into a solution of its ion at unit activity develops a potential relative to the standard hydrogen electrode called the standard electrode potential of the metal. This is negative for electropositive metals because they tend to go into solution as ions, leaving electrons behind on the electrode. An inert electrode dipped into a solution of an oxidizing couple also develops a potential known as the standard electrode potential of the couple.

Electrode potentials, and the various equations containing electrode potentials, always refer to the half reaction written with the electrons on the left of the equation, i.e. the reduction half reaction.

Electrode potentials are intensive properties and cannot be added together. Reactions can be combined if potentials are converted to free energies by the relationship:

$$\Delta G = -nFE$$

where ΔG is the free energy change for the half reaction concerned and n is the number of electrons in the half reaction.

Electrode potentials under non-standard conditions are related to standard electrode potentials by the Nernst equation:

$$E = E^{\ominus} + \frac{RT}{nF} \ln\left(\frac{[\text{oxidized species}]}{[\text{reduced species}]}\right)$$

$$= E^{\ominus} + \frac{0.0591}{n} \log\left(\frac{[\text{oxidized species}]}{[\text{reduced species}]}\right) \text{ at } 25\,^{\circ}\text{C}$$

Standard electrode potentials can be used to calculate equilibrium constants via the relationships

and

$$\Delta G^{\ominus} = nF^{\ominus}E^{\ominus}$$

$$\Delta G^{\ominus} = -2.303\,RT \log K$$

Appendix

Standard Electrode Potentials

	Half reaction	$E^{\ominus}/V$
$Li^+ + e^-$	$= Li$	-3.04
$Ca^{2+} + 2e^-$	$= Ca$	-2.87
$Na^+ + e^-$	$= Na$	-2.71
$Mg^{2+} + 2e^-$	$= Mg$	-2.37
$Al^{3+} + 3e^-$	$= Al$	-1.66
$Mn^{2+} + 2e^-$	$= Mn$	-1.18
$Zn^{2+} + 2e^-$	$= Zn$	-0.76
$Cr^{3+} + 3e^-$	$= Cr$	-0.74
$Fe^{2+} + 2e^-$	$= Fe$	-0.44
$Cr^{3+} + e^-$	$= Cr^{2+}$	-0.41
$Co^{2+} + 2e^-$	$= Co$	-0.28
$N_2 + 5H^+ + 4e^-$	$= N_2H_5^+$	-0.23
$H^+ + e^-$	$= \frac{1}{2}H_2$	0
$S + 2H^+ + 2e^-$	$= H_2S$	$+0.14$
$Sn^{4+} + 2e^-$	$= Sn^{2+}$	$+0.15$
$Cu^{2+} + e^-$	$= Cu^+$	$+0.15$
$\frac{1}{2}N_2 + 4H^+ + 3e^-$	$= NH_4^+$	$+0.27$
$Cu^{2+} + 2e^-$	$= Cu$	$+0.34$
$Cu^+ + e^-$	$= Cu$	$+0.52$
$\frac{1}{2}I_2 + e^-$	$= I^-$	$+0.54$
$MnO_4^- + e^-$	$= MnO_4^{2-}$	$+0.56$
$O_2 + 2H^+ + 2e^-$	$= H_2O_2$	$+0.68$
$Fe^{3+} + e^-$	$= Fe^{2+}$	$+0.76$
$Ag^+ + e^-$	$= Ag$	$+0.80$
$Hg^{2+} + e^-$	$= \frac{1}{2}Hg_2^{2+}$	$+0.92$
$\frac{1}{2}Br_2 + e^-$	$= Br^-$	$+1.07$
$IO_3^- + 6H^+ + 5e^-$	$= \frac{1}{2}I_2 + 3H_2O$	$+1.19$
$\frac{1}{2}O_2 + 2H^+ + 2e^-$	$= H_2O$	$+1.23$
$Cr_2O_7^{2-} + 14H^+ + 6e^-$	$= 2Cr^{3+} + 7H_2O$	$+1.33$
$\frac{1}{2}Cl_2 + e^-$	$= Cl^-$	$+1.36$
$Ce^{4+} + e^-$	$= Ce^{3+}$	$+1.45$
$Mn^{3+} + e^-$	$= Mn^{2+}$	$+1.50$
$MnO_4^- + 8H^+ + 5e^-$	$= Mn^{2+} + 4H_2O$	$+1.52$
$BrO_3^- + 6H^+ + 5e^-$	$= \frac{1}{2}Br_2 + 3H_2O$	$+1.52$
$MnO_4^- + 4H^+ + 3e^-$	$= MnO_2 + 2H_2O$	$+1.69$
$Pb^{4+} + 2e^-$	$= Pb^{2+}$	$+1.70$
$Co^{3+} + e^-$	$= Co^{2+}$	$+1.82$
$S_2O_8^{2-} + 2e^-$	$= 2SO_4^{2-}$	$+2.01$
$\frac{1}{2}F_2 + e^-$	$= F^-$	$+2.80$

Programme 5

Oxidation State Diagrams

Objectives

After completing this programme, you should be able to:
1. Plot the oxidation state diagram of an element, given electrode potential data.
2. Use the diagram to obtain $E^{\ominus}$ for any ion-electron half reaction represented.
3. Comment on the effects of changes of conditions on the solution chemistry of the element concerned.
4. Use the diagram to comment on general features of the chemistry of the element concerned, including predicting the occurrence of disproportionation of a state, or the occurrence of oxidation–reduction reactions between different states of the same element.

All four objectives are tested at the end of the programme.

Assumed Knowledge

Before starting the programme you should have a knowledge of oxidation numbers, standard electrode potential ($E^{\ominus}$) and standard Gibbs free energy change ($\Delta G^{\ominus}$). You should understand and be able to use the expression $\Delta G^{\ominus} = -nFE^{\ominus}$ and you should be able to calculate $E^{\ominus}$ for a reaction, given $E^{\ominus}$ for two related reactions. These topics are the subject of earlier programmes in this book.

Oxidation State Diagrams

5.1   In programme 4 we met the expression

$$\Delta G^{\ominus} = -nFE^{\ominus}$$

This relates the standard free energy change, $\Delta G^{\ominus}$, to the standard potential, $E^{\ominus}$, for an ion-electron half reaction involving n electrons. F is the Faraday ($= 96\,487\ C\,mol^{-1}$).

Use this relationship to complete the following table:

Half reaction	$E^{\ominus}/V$	$\Delta G^{\ominus}/J\,mol^{-1}$
$Ti^{4+} + e^- \rightarrow Ti^{3+}$	$+0.06$	$-0.06F$
$Ti^{3+} + e^- \rightarrow Ti^{2+}$	-0.37	
$Ti^{2+} + 2e^- \rightarrow Ti$	-1.63	

5.2

Half reaction	$E^{\ominus}/V$	$\Delta G^{\ominus}/J\,mol^{-1}$
$Ti^{4+} + e^- \rightarrow Ti^{3+}$	$+0.06$	$-0.06F$
$Ti^{3+} + e^- \rightarrow Ti^{2+}$	-0.37	$+0.37F$
$Ti^{2+} + 2e^- \rightarrow Ti$	-1.63	$+3.26F$

(Note that for many applications of these ideas we do not need to use the actual value of the Faraday, F).

We can use the data in the table above to calculate the value of $E^{\ominus}$ for any other half reaction, e.g.

$$Ti^{3+} + 3e^- \rightarrow Ti$$

This equation is easily obtained by adding up two of the half reactions in the table.

Write down these equations and add them up to obtain this final result.

5.3 $Ti^{3+} + e^- \rightarrow Ti^{2+}$ $\Delta G^{\ominus} = +0.37\,F$
        $Ti^{2+} + 2e^- \rightarrow Ti$ $\Delta G^{\ominus} = +3.26\,F$
        _____

        $Ti^{3+} + 3e^- \rightarrow Ti$
        _____

We must remember now that standard potentials are intensive properties (like density or temperature) and cannot be added directly. We
can, however, add the standard free energy changes because these are
extensive properties.

What is the standard free energy change for the reaction

$$Ti^{3+} + 3e^- \rightarrow Ti$$

5.4 $Ti^{3+} + 3e^- \rightarrow Ti$ $\Delta G^{\ominus} = 0.37\,F + 3.26\,F = 3.63\,F$
Use the relationship $\Delta G^{\ominus} = -nFE^{\ominus}$ to calculate $E^{\ominus}$ from this free
energy value.

5.5 $E^{\ominus} = -3.63\,F/3\,F = -1.21$ V.

Clearly this value is not simply the sum of the two potentials concerned, it is in fact the *weighted mean* of the two values, i.e.

$$\frac{(-1.63 \times 2) - (0.37 \times 1)}{3}$$

Using the weighted mean technique for $E^{\ominus}$ calculations, however,
can lead to confusion in cases where some equations (and hence the
sign of $E^{\ominus}$ values) are reversed; the slightly more laborious $\Delta G^{\ominus}$
method is recommended.

Calculate the standard potential for the half reaction

$$Ti^{4+} + 2e^- \rightarrow Ti^{2+}$$

5.6 $Ti^{4+} + e^- \rightarrow Ti^{3+}$ $\Delta G^\ominus = -0.06\,F$
         $Ti^{3+} + e^- \rightarrow Ti^{2+}$ $\Delta G^\ominus = +0.37\,F$

         $Ti^{4+} + 2e^- \rightarrow Ti^{2+}$ $\Delta G^\ominus = +0.31\,F$ $E^\ominus = -0.155$ V

All the relationships we have used and calculated so far can be shown graphically. Although graphical methods are less precise than numerical ones for calculations, the result does provide, in pictorial form, a considerable amount of useful information about the chemistry of the element concerned.

The method simply involves plotting the oxidation state along the x axis (horizontal) and $nE^\ominus$ along the y axis (vertical), taking the origin to be the element itself in the zero oxidation state. The resulting diagram is a series of lines whose slopes are equal to the $E^\ominus$ value between the oxidation states considered, as can be seen from the rather self-evident equation:

$$nE^\ominus = E^\ominus n$$

or

$$y = mx$$

Plot the point for Ti^{2+} on the axes given:

$Ti^{2+} + 2e^- \rightarrow Ti$ $E^\ominus = -1.63$ V
Since two electrons are involved, $nE^\ominus = -3.26$ V.

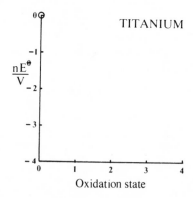

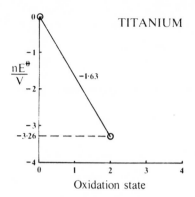

In this diagram, the point for Ti^{2+} has been plotted at $nE^{\ominus} = -3.26$ V and the value of the slope of the line $(= E^{\ominus})$ has also been included. The point for Ti^{3+} can now be added at a position such that the slope of the line from Ti^{2+} to Ti^{3+} equals the potential

$$Ti^{3+} + e^{-} \rightarrow Ti^{2+} \qquad E^{\ominus} = -0.37 \text{ V}$$

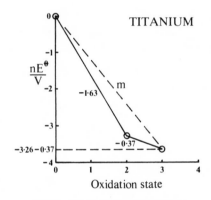

Find the slope, m, of the line between the zero and $+3$ states, and compare it with the potential of -1.21 V given in Frame 5.5.

5.8 $m = (-3.26 - 0.37)/3 = -1.21$ V, so the diagram still has the properties claimed for it, that the slope of the line between two points equals the standard potential for the half reaction concerned.

Finally, plot the point for Ti^{4+}, using the potential

$$Ti^{4+} + e^{-} \rightarrow Ti^{3+} \qquad E^{\ominus} = +0.06 \text{ V}$$

and show that the point agrees with the potential

$$Ti^{4+} + 4e^{-} \rightarrow Ti \qquad E^{\ominus} = -0.8925 \text{ V}$$

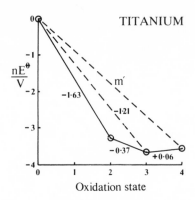

The point for Ti^{4+} should be plotted at $nE^{\ominus}/V = -3.57$, i.e. $(-2 \times 1.63) - 0.37 + 0.06$. The slope m' is then $-3.57/4 = -0.8925$ V.

We can now summarize the properties of an oxidation state diagram as follows:

i. One point is at the origin (the element in the zero oxidation state).
ii. The horizontal axis is the oxidation state of the element.
iii. The slope of the line between any two points equals $E^{\ominus}$ between those two oxidation states.

Plot a diagram for copper (the diagram looks completely different from that for titanium!) using the data below. The first point, at the origin, has been drawn in already.

$$Cu^{2+} + 2e^{-} \rightarrow Cu \qquad E^{\ominus} = +0.34 \text{ V}$$
$$Cu^{+} + e^{-} \rightarrow Cu \qquad E^{\ominus} = +0.52 \text{ V}$$

COPPER

Oxidation state

5.10

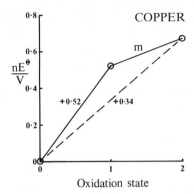

COPPER

Use the diagram to find the potential for

$$Cu^{2+} + e^- = Cu^+ \qquad E^{\ominus} = m$$

5.11 $m = +0.16$ V, in agreement with the result which could have been calculated via standard free energies.

So far, oxidation state diagrams have been approached from a rather empirical point of view, but we shall now look more deeply at the significance of the vertical axis. Since standard electrode potential and free energy are related by the expression $\Delta G^{\ominus} = -nFE^{\ominus}$, and since F is a constant, $nE^{\ominus}$ is a measure of the negative free energy change for the reaction.

Higher oxidation state + electrons→lower oxidation state

What is the effect on the free energy change of reversing the direction of a reaction?

5.12   The free energy change reverses its sign. If, therefore, we consider the reaction

Low oxidation state→higher oxidation state + electrons

the quantity $nE^{\ominus}$ is a measure of standard free energy change.

Since the horizontal axis of our diagram proceeds in this direction, from lower to higher oxidation state, we can interpret the vertical axis as a direct measure of the standard free energy of each oxidation state.

It must be recalled at this point that the free energy change is relative to the $2H^+/H_2$ couple, but the results in practice are a useful reflection of the aqueous chemistry of the element concerned.

Calculate $\Delta G^{\ominus}$ in kJ mol^{-1} for the change $Cu \rightarrow Cu^{2+} + 2e^-$, given that $Cu^{2+} + 2e^- \rightarrow Cu$ $E^{\ominus} = +0.34$ V ($\equiv$ J C^{-1}). $F = 96\,487$ C mol^{-1}.

5.13   For $Cu^{2+} + 2e^- \rightarrow Cu$ $\Delta G^{\ominus} = -2 \times 96\,487 \times 0.34 \text{ J mol}^{-1}$
$$= -65.6 \text{ kJ mol}^{-1}$$
Thus for $Cu \rightarrow Cu^{2+} + 2e^-$ $\Delta G^{\ominus} = +65.6 \text{ kJ mol}^{-1}$
(relative to the hydrogen couple), i.e.

$$Cu \rightarrow Cu^{2+} + 2e^-$$
$$2e^- + 2H^+ \rightarrow H_2$$

$Cu + 2H^+ \rightarrow Cu^{2+} + H_2$ $\Delta G^{\ominus} = +65.6 \text{ kJ mol}^{-1}$

This last equation is thermodynamically correct. It indicates that metallic copper does not displace hydrogen from an acid, i.e. copper is less electropositive than hydrogen. Hence, on the oxidation state diagram for copper, the point for Cu^{2+} appears above that for the metal.

We can increase the usefulness of the diagrams if we include a second vertical axis (on the right of the diagram) showing standard free energy values. Although we must always remember that these are relative to hydrogen, they do, like the $nE^{\ominus}$ values, give a useful indication of the ordinary aqueous chemistry of the element concerned.

Calculate the value of $\Delta G^{\ominus}$ for a 1-electron change in which $E^{\ominus} = 1$ V ($F = 96\,487$ C mol^{-1}).

5.14 $-96.487 \text{ kJ mol}^{-1}$.

Since $\Delta G^{\ominus} = -nFE^{\ominus}$, if $nE^{\ominus} = 1 \text{ J C}^{-1}$, then

$$\Delta G^{\ominus} = \frac{-96\,487}{1000} \text{ C mol}^{-1} \text{kJ C}^{-1}$$

$$= -96.487 \text{ kJ mol}^{-1}$$

or $+96.487 \text{ kJ mol}^{-1}$ for the reverse reaction. Thus, each unit of $nE^{\ominus}$ corresponds to $96.487 \text{ kJ mol}^{-1}$.

With this extra information included, the diagrams we have so far constructed appear as follows:

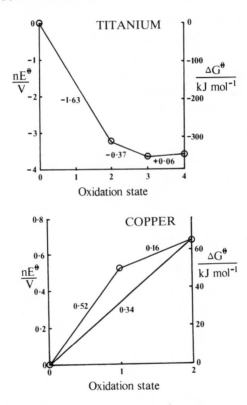

From your knowledge that systems tend to achieve a minimum free energy, predict the most stable oxidation states of copper and titanium from the two diagrams.

5.15   The most stable oxidation states are copper zero and, titanium $+3$ (since these states are the ones in which the free energy is the lowest).

This tendency of the system to 'run downhill' can be a useful guide to the general chemistry of an element, given the oxidation state diagram.

What, for instance, can we say about the likely oxidizing or reducing properties of Ti^{2+}?

5.16   It is likely to be a reducing agent, being readily oxidized up to Ti^{3+} or to Ti^{4+}, both of which have lower standard free energies.

Similarly, what can we say about the reactivity of metallic copper towards oxidizing agents?

5.17   It is likely to be rather inert unless the oxidizing agent is fairly powerful. Both of these simple predictions are observed in practice in the chemistry of these elements.

We shall turn now to some more precise information which the diagrams can provide. Consider a mixture of 0.5 mol of Cu metal and 0.5 mol of Cu^{2+}. The standard free energy of Cu metal is taken to be zero and that of Cu^{2+} is $+65.6$ kJ mol^{-1}. What, then, is the standard free energy of the mixture?

5.18   32.8 kJ mol^{-1}, i.e. zero for the Cu metal and 0.5×65.6 for the Cu^{2+}.

The average oxidation state of the equimolar mixture of Cu(0) and Cu(II) is clearly $+1$, so we can now plot the position of the mixture on an oxidation state diagram. Do this, using the diagram in Frame 5.14.

5.19

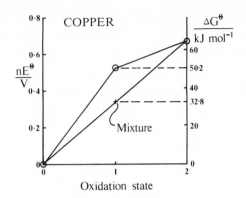

Oxidation state

i.e. the mixture lies directly beneath the Cu^+ point on the diagram.

Thus the mixture of $\frac{1}{2}Cu + \frac{1}{2}Cu^{2+}$ has a standard free energy of $+32.8 \ kJ \ mol^{-1}$, while Cu^+ has a standard free energy of $+50.2 \ kJ \ mol^{-1}$.

What, then, is the standard free energy change of the reaction

$$Cu^+ \rightarrow \frac{1}{2}Cu + \frac{1}{2}Cu^{2+} \tag{5.1}$$

5.20 $\Delta G^{\ominus} = G^{\ominus}$ (products) $- G^{\ominus}$ (reactants)
       $= 32.8 \qquad\qquad - 50.2$
       $= -17.4 \ kJ \ mol^{-1}$

What can you say about the spontaneity, or otherwise, of this reaction?

5.21   It should occur spontaneously, i.e. the $+1$ oxidation state of copper is predicted to disproportionate into the zero and $+2$ states. The standard free energy change of the disproportionation reaction is $-17.4 \text{ kJ mol}^{-1}$. It is not necessary to go to the trouble of calculating standard free energies in order to predict disproportionation reactions. Any oxidation state which lies on a point which is convex when viewed from above [e.g. Cu(I)] is liable to disporportionate. Any oxidation state lying on a point which is concave from above [e.g. Ti(II) or Ti(III)] will not undergo disproportionation.

Note that the oxidation state diagram has the limitation of any thermodynamic treatment in that it tells us nothing of how fast the disproportionation will occur. In some cases there may also be more than one disproportionation route, and again the diagram gives no information about which will be kinetically favoured.

The oxidation state diagram for manganese is given below. Identify the two oxidation states which lie on convex points and hence are unstable to disproportionation.

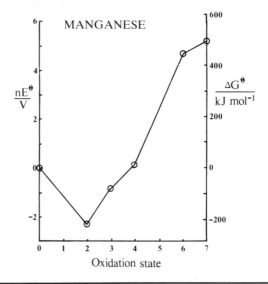

5.22   The $+3$ and $+6$ states lie on convex points. Hence, Mn(III) and Mn(VI) are unstable to disproportionation. They are found to react as follows:

$$2H_2O + 2Mn^{3+} \rightarrow MnO_2\downarrow + Mn^{2+} + 4H^+ \qquad (5.2)$$

i.e. disproportionation of Mn(III) to Mn(IV) $+$ Mn(II), and

$$3MnO_4^{2-} + 4H^+ \rightarrow MnO_2\downarrow + 2MnO_4^- + 2H_2O \qquad (5.3)$$

i.e. disproportionation of Mn(VI) to Mn(IV) $+$ Mn(VII).

Note that Mn(VI) is also unstable with respect to disproportionation into Mn(II) and Mn(VII), although the free energy change is greatest for the formation of MnO_2.

Oxidation state diagrams can also be used to predict the occurrence of reactions which are the reverse of disproportionation, namely the oxidation of a low state by a higher state of the same element. This type of reaction can be shown by a procedure exactly like that used to indicate disproportionation in copper in Frames 5.17–5.21.

Consider, for example, a mixture of 3 mol of Mn(II) and 2 mol of Mn(VII). What is the average oxidation state of the 5 mol making up this mixture?

5.23   Average oxidation state $= +4$, i.e. 3 mol of Mn(II) $+ 2$ mol of Mn(VII):

$$\text{Mean} = \frac{(3 \times 2) + (2 \times 7)}{5} = +4$$

The position of the mixture on the oxidation state diagram is therefore on the line joining the oxidation states $+2$ and $+7$, which lies slightly above the point representing Mn(IV).

Since the mixture lies above the $+4$ state on the free energy diagram, how could it change chemically in order to decrease its free energy?

5.24   It could self-oxidize and reduce to form Mn(IV), i.e.

$$2MnO_4^- + 3Mn^{2+} + 2H_2O \rightarrow 5MnO_2\downarrow + 4H^+$$

Again, it is probably clear to you that if the line between any two oxidation states passes above the point representing an intermediate state, then the highest state will oxidize the lowest to the intermediate one. This, incidentally, is the reason why MnO_4^- goes in the burette in a titration; if it were placed in the flask, any Mn^{2+} formed by reduction would immediately be oxidized to a dark precipitate of manganese dioxide. Titrations involving manganese(VII) must be arranged to keep Mn(II) and Mn(VII) away from each other as much as possible.

Look back at the diagram for titanium in Frame 5.14 and see if there are two states which can be joined by a line which then passes above an intermediate oxidation state. (*N.B.* do not forget the zero state).

5.25   The line joining Ti(IV) to Ti(II) passes above Ti(III).
        The line joining Ti(IV) to Ti passes above Ti(II) and Ti(III).
        The line joining Ti(III) to Ti passes above Ti(II).

Hence in all these cases, the intermediate oxidation state is predicted to be more stable than the mixture, i.e. for the first example, reaction 5.4 should proceed as written:

$$Ti(IV) + Ti(II) \rightarrow 2Ti(III) \qquad (5.4)$$

Write the corresponding equations for the other changes indicated above.

5.26 $$Ti(IV) + Ti \rightarrow 2Ti(II) \qquad (5.5)$$
        $$3Ti(IV) + Ti \rightarrow 4Ti(III) \qquad (5.6)$$
        $$2Ti(III) + Ti \rightarrow 3Ti(II) \qquad (5.7)$$

These equations can be balanced by using the techniques discussed in Programme 2.

We can, again, obtain our information direct from the diagram since the points corresponding to Ti(II) and Ti(III) are concave when viewed from above. This indicates that they are stable to disproportionation and, indeed, should be capable of being the products of reactions which are the reverse of disproportionations.

Look back at the diagram for manganese in Frame 5.21. Which oxidation state of manganese has the lowest standard free energy and should therefore be the most stable state of the element?

5.27   Mn(II).

Could this state be formed by a 'reverse disproportionation' process and, if so, from what other two oxidation states?

5.28   It should be possible to form Mn(II) from manganese metal and any of the other states since the lines joining Mn to the other states pass above the Mn^{2+} point. These reactions may, however, be slow in many instances.

So far, the diagrams plotted have referred to standard electrode potentials, when every reactant and every product is at unit activity and, in particular, when hydrogen ions are present at unit activity. It is equally possible to plot diagrams for different conditions and much useful information can be obtained from the way in which the diagrams change as the conditions are altered. In alkaline solution, for instance, the manganese diagram changes as shown below.

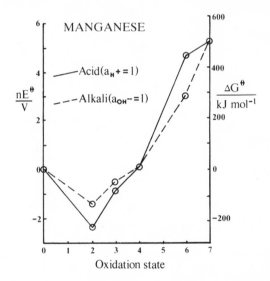

Which oxidation state of Mn is on a convex point (unstable to disproportionation) in acidic solution, but not in alkaline solution?

5.29   Mn(VI). Hence MnO_4^{2-} is unstable to disproportionation in acidic but not in alkaline solution.

In fact, it is possible to prepare manganate(VII) (i.e. permanganate) by oxidizing manganese dioxide in a strongly alkaline solution to manganate(VI), and then acidifying to cause disproportionation of manganate(VI) to manganate(VII) and manganate(IV). These changes can easily be understood by reference to the oxidation state diagram.

Plot an oxidation state diagram for iron to show the effect of complexing iron(II) and iron(III) with water, cyanide and 1,10-phenanthroline (phen), given the following potentials in volts. The vertical axis needs to go from zero down to -2.5 and the horizontal axis from zero to 3.

Couple	Aquo	Cyano	Phen
$Fe^{2+} + 2e^- \rightarrow Fe$	-0.44	-1.15	-1.06
$Fe^{3+} + 3e^- \rightarrow Fe$	-0.04	-0.65	-0.32

5.30

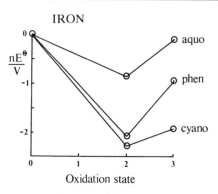

The above diagram clearly shows the considerable stabilization achieved by using the ligands, and also the fact that the relative stability of iron(II) and iron(III) is markedly dependent on the ligand. The potential between these two states in the presence of 1,10-phenanthroline makes the system very useful as an indicator in redox titrations, where it is known as ferroin:

$$Fe(phen)_3^{2+} \quad \rightleftharpoons \quad Fe(phen)_3^{3+} + e^-$$

reducing agent   oxidizing agent
present   present
deep red   very pale blue

Further Reading

1. A. A. Frost, *J. Am. Chem. Soc.*, 1951, **73**, 2680. (The first proposal of this form of presentation. Some of the symbols and units are now outdated, and $E^{\ominus}$ values are quoted on the old American convention with opposite signs to the IUPAC system.)
2. E. A. V. Ebsworth, *Educ. Chem.*, 1964, **1**, 123. (A 'rediscovery' of the diagrams—he had clearly not seen the paper by Frost. He quotes several diagrams, including all those for the first-row transition metals Ti to Zn, and discusses clearly the properties of the diagrams.)
3. C. S. G. Phillips and R. J. P. Williams, *Inorganic Chemistry,* Vol. 1, Oxford University Press, London, 1965, p. 314. (The diagrams are discussed in Vol. 1, using nitrogen as an example, and are used throughout both volumes of the book.)
4. R. V. Parish, *The Metallic Elements,* Longmans, London, 1977. (The properties of the diagrams are clearly explained and they are used when discussing the chemistry of individual elements.)

Oxidation State Diagrams Test

Question numbers correspond to the programme objectives.

1. Plot an oxidation state diagram for thallium, showing the effect of perchloric acid and hydrochloric acid solution, using the following potential values in volts:

Couple	$HClO_4$ soln.	HCl soln.
$Tl^+ + e^- = Tl$	-0.34	-0.55
$Tl^{3+} + 2e^- = Tl^+$	$+1.25$	$+0.77$

2. Use your diagram for thallium to find $E^{\ominus}$ for Tl^{3+}/Tl in HCl and in $HClO_4$.
3. a. Which of the anions (Cl^- or ClO_4^-) most assists the formation of thallium ions in solution?
   b. Which of the two oxidation states of thallium is the more stabilized?
4. A diagram for nitrogen is given below. Use this diagram to comment briefly on the likely stability of the substances given. If you think a substance is unstable, suggest possible decomposition products:

N_2
NH_4NO_3 [contains N($-$III) and N(V)]
NH_4NO_2 [contains N($-$III) and N(III)]
H_2NOH

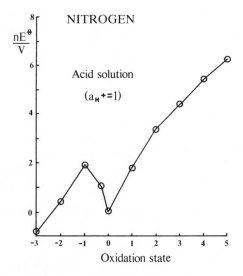

Answers

1.

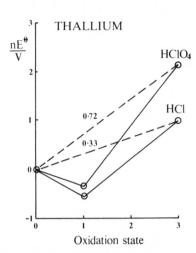

The points should lie at the following values of $nE^{\ominus}/V$:

-0.34	Tl(I)	$HClO_4$	2
-0.55	Tl(I)	HCl	2
$+2.16$	Tl(III)	$HClO_4$	2
$+0.99$	Tl(III)	HCl	2

2. $E^{\ominus}$, Tl(III)/Tl in HCl $= 0.99/3 = 0.33$ V.   2

$E^{\ominus}$, Tl(III)/Tl in $HClO_4 = 2.16/3 = 0.72$ V.   2

These values are given by the slopes of the dashed lines on the diagram.

3. a. Cl^-, because the points lie below those for ClO_4^- on the vertical (free energy) axis.   2

 b. Tl(III), because the gap between the perchlorate and chloride points is greater for Tl(III) than for Tl(I).   2

4. N_2 Very stable, occurs at a deep minimum on the diagram. No chance of disproportionation.   2

NH_4NO_3 NO_3^- [nitrogen (V)] should be able to oxidize NH_4^+ [nitrogen ($-$III)] to a number of intermediate oxidation states. In fact, the compound decomposes on heating to produce $N_2O + 2H_2O$.   2

NH_4NO_2 Similar to ammonium nitrate. The line joining NH_4^+ and NO_2^- passes above several other compounds. In fact, it decomposes on heating to produce mainly $N_2 + 2H_2O$ 2

H_2NOH Hydroxylamine should be liable to disproportionate since it is on a convex point. In fact, it is very unstable and disproportionates to N_2, N_2O and NH_3.   2

Total   24

The four test questions are designed to test different aspects of your understanding of this topic. An overall score of 19 or more indicates a good grasp of the subject.

Oxidation State Diagrams

Revision Notes

An oxidation state diagram is a graph of $nE^{\ominus}$ (y axis) against oxidation state (x axis) for an element, where n is the oxidation state change and $E^{\ominus}$ is the standard potential of the state relative to the free element.

Oxidation state diagrams are completely defined by three properties:
  i.   One point is at the origin (the element in the zero oxidation state).
  ii. The horizontal axis is the oxidation state of the element.
  iii. The slope of the line between any two points equals $E^{\ominus}$ between those two oxidation states.

The diagrams give a good general picture of the chemistry of the element concerned, since the vertical axis is a measure of free energy. Thus any minimum or concave point on the diagram represents a state which is stable relative to those near it, and any maximum or convex point represents a state which is unstable to disproportionation into two other oxidation states. If the line joining a high to a low oxidation state passes above the point representing some intermediate state, then the high state should be thermodynamically capable of oxidizing the low state. The effect of different solution conditions (pH, ligands, etc.) can be represented on the diagram by plotting potentials under the appropriate conditions.

Index

NOW AVAILABLE

To illustrate the concepts of oxidation state diagrams referred to in Programme 5, the author has prepared a disc containing two computer programs to allow students to extend their practice in this area. One program allows oxidation state diagrams or tables of data to be presented. The other program constructs an oxidation state diagram from the user's own data.

The 5.25 inch disc is currently available for the British Broadcasting Corporation (BBC) Microcomputer Model B with a 40 track disc drive. Copies can be ordered through your bookseller or direct from John Wiley & Sons Ltd using the order form below, price £9.95 (+ VAT), $14.50 each.

A full set of disc operating instructions is included with each disc.

Vincent: **Oxidation and Reduction**—Supplementary disc

Please send me. . . . . copies of the Vincent: **Oxidation and Reduction** supplementary disc 0 471 90765 0 at £9.95 (+ VAT), $14.50 each.

Please send me more information about the disc.

I would be interested in buying the disc if this was available for the . . . . machine/s.

POSTAGE AND HANDLING FREE FOR CASH WITH ORDER OR PAYMENT BY CREDIT CARD
☐ Remittance enclosed. . . . . .Allow approx. 14 days for delivery
☐ Please charge this order to my credit card (All orders subject to credit approval)
Delete as necessary:—AMERICAN EXPRESS, DINERS CLUB, BARCLAYCARD/VISA, ACCESS

CARD NUMBER ☐☐☐☐☐☐☐☐☐☐☐☐☐☐☐☐☐☐☐ Expiration date . . . .
☐ Please send me an invoice for prepayment. A small postage and handling charge will be made.
Software purchased for professional purposes is generally recognized as tax deductible.
☐ Please keep me informed of new books in my subject area which is . . . .

NAME/ADDRESS . . . .

. . . .

. . . .

OFFICIAL ORDER No. . . . . SIGNATURE . . . .

**ENHANCE THE INTERACTIVE NATURE OF YOUR
PROGRAMMED LEARNING TEXT**

Order your discs through your bookseller or direct from Wiley:

Dr P T Shepherd, Chemistry Editor,
John Wiley & Sons, Ltd.,
Baffins Lane, Chichester,
Sussex PO19 1UD UNITED KINGDOM

Vincent—**Oxidation and Reduction**—Supplementary disc

Affix
Postage
Stamp

Dr P T Shepherd, Chemistry Editor,
John Wiley & Sons, Ltd.,
Baffins Lane, Chichester,
Sussex PO19 1UD UNITED KINGDOM